FORGING A LABORING RACE

CULTURE, LABOR, HISTORY SERIES
General Editors: Daniel Bender and Kimberley L. Phillips

*Working the Diaspora: The Impact of African Labor on the
Anglo-American World, 1650–1850*
Frederick C. Knight

*Class Unknown: Undercover Investigations of American Work and
Poverty from the Progressive Era to the Present*
Mark Pittenger

*Steel Barrio: The Great Mexican Migration to South Chicago,
1915–1940*
Michael D. Innis-Jiménez

*Fueling the Gilded Age: Railroads, Miners, and Disorder in
Pennsylvania Coal Country*
Andrew B. Arnold

*A Great Conspiracy against Our Race: Italian Immigrant Newspapers
and the Construction of Whiteness in the Early 20th Century*
Peter G. Vellon

*Reframing Randolph: Labor, Black Freedom, and the Legacies of
A. Philip Randolph*
Edited by Andrew E. Kersten and Clarence Lang

Making the Empire Work: Labor and United States Imperialism
Edited by Daniel E. Bender and Jana K. Lipman

*Whose Harlem Is This, Anyway? Community Politics and
Grassroots Activism during the New Negro Era*
Shannon King

*Health in the City: Race, Poverty, and the Negotiation of Women's
Health in New York City, 1915–1930*
Tanya Hart

*Trotskyists on Trial: Free Speech and Political Persecution since
the Age of FDR*
Donna T. Haverty-Stacke

*Forging a Laboring Race: The African American Worker in the
Progressive Imagination*
Paul R. D. Lawrie

Forging a Laboring Race

The African American Worker in the
Progressive Imagination

Paul R. D. Lawrie

NEW YORK UNIVERSITY PRESS
New York

NEW YORK UNIVERSITY PRESS
New York
www.nyupress.org

References to Internet websites (URLs) were accurate at the time of writing. Neither the author nor New York University Press is responsible for URLs that may have expired or changed since the manuscript was prepared.

Library of Congress Cataloging-in-Publication Data
Names: Lawrie, Paul R. D., author.
Title: Forging a laboring race : the African American worker in the Progressive imagination / Paul R.D. Lawrie.
Description: New York : New York University Press, 2016. | Series: Culture, labor, history series | Includes bibliographical references and index.
Identifiers: LCCN 2016001632 | ISBN 9781479857326 (cl : alk. paper)
Subjects: LCSH: African Americans—History—1877-1964. | African Americans—Employment—History—20th century. | Working class African Americans—History—20th century. | Labor—United States—History—20th century. | Industrialization—United States—History—20th century. | United States—Race relations—History—20th century.
Classification: LCC E185.6 .L345 2016 | DDC 331.6/396073—dc23
LC record available at http://lccn.loc.gov/2016001632

New York University Press books are printed on acid-free paper, and their binding materials are chosen for strength and durability. We strive to use environmentally responsible suppliers and materials to the greatest extent possible in publishing our books.

Manufactured in the United States of America

10 9 8 7 6 5 4 3 2 1

Also available as an ebook

For my Children

William Anthony Ellison Lawrie

Zoë Madeline Lawrie

And

As Always To Rose

CONTENTS

ACKNOWLEDGMENTS

The best part about completing this book is having the chance to thank the various people and institutions that made it possible. My debts are numerous. This project was brought to life by an amazing set of archivists who answered my vague and rambling inquiries with patience, humor, and unstinting professionalism. I would like to thank the staff at the Library of Congress in Washington, D.C., and the Schomburg Centre for Research in Black Culture in New York; Mary Ann Quinn at the Rockefeller Archive Centre; Fath Ruffins at the Smithsonian Institution in Washington, D.C.; Earl Spamer at the American Philosophical Society in Pennsylvania; and Richard Boylen and Walter Hill at the National Archives in College Park, Maryland. Mr. Hill's immense knowledge of the National Archives and Records Administration's holdings in African American history saved me innumerable hours and led me on many productive detours. Even in the last stages of a long illness, Mr. Hill remained a model of dignity and grace.

Initial research was facilitated by grants from the Department of History, University of Toronto; Department of Humanities, University of Toronto Scarborough; and the Centre for American Studies (CSUS) at Trinity College, University of Toronto. Further assistance was provided by grants from the Rockefeller Archives Centre in New York and a Library Fellowship at the American Philosophical Society in Philadelphia. Final funds were provided by a Major Research Grant, Travel Grant, and SEED funding from the Faculty of Arts at the University of Winnipeg. Many thanks to Clara Platter, Dorothea Stillman Halliday, and Constance Grady at NYU Press for guiding me through the long and often confusing publishing processes. Thanks to the anonymous readers for NYU and to Jodi Narde for her rigorous and expert editing of the final manuscript. Thanks to Daniel Bender and Kimberley Phillips for including this book in the Culture, Labor, History series. Parts of chapter 1 appeared previously as "'Mortality as the Life Story of a People': Frederick L. Hoffman and Actuarial Narratives of African American Extinction, 1896–1915" in the *Canadian Review of American Studies* 43, no.3 (2013) and I am grateful for the permission to make use of that material.

Over the years I received invaluable guidance from a diverse group of scholars. Thanks to Stephen Brooke and Marc Stein at York University for providing wonderful models of engaged and rigorous scholarship and setting a young undergraduate on the path—for better or worse—to a life in academia. At the University of Toronto I would like to thank Elspeth Brown, Russ Kazal, Yonatan Eyal, Kenneth Bartlett, and Randall Hansen for their support. Most importantly, I was fortunate to have a remarkably supportive and gifted set of advisors who enhanced my graduate school experience every step of the way. Michael "discourse and paradigms" Wayne provided much-needed intellectual and personal perspective in navigating graduate school. I am forever grateful to Rick Halpern who fought for and supported me in every possible way, from my first day of graduate school to the very last. Daniel Bender was a model advisor, providing any and all forms of intellectual, financial, and emotional support. I simply could not have asked for a better mentor and friend.

A wonderful cohort of my peers helped relieve much of the isolation, tedium, and anxiety of graduate school and shaped the project in unintended and unexpected ways. Thanks to Nancy Catton, Benjamin Pottruff, Nathan Cardon, Camille Begin, Nadia Jones Gailani, Holly Karibo, Ian Rocksborough-Smith, and Jared Toney. My friends and family outside the program sustained and supported me throughout the years. Thank you to my best man Damian Temporale and his partner, Tina Burke, whose quiet strength and compassion for their wonderful family is an inspiration. To old friends, Matthew Nailer, Mike Mjanes, Ian Gooley, Sandra and Brian Neary, Patricia Moretti, and Natalie Boccia—and to Michael Hersh for his friendship and hospitality during many research trips to NYC. To Dr. Jeremy Shtern, my co-conspirator in academia and baseball analysis, an excellent and committed scholar, confidant, and dear friend. Very special thanks to Davarian Baldwin, for profound and consistent insights and for being a model of scholarly rigor, dedication, and collegiality, and a testament that exceptional scholarship can and should matter within and beyond the ivory tower.

I would also like to extend my thanks to all my colleagues at the University of Winnipeg, especially those in the Department of History for creating a collegial environment in which to shepherd this project through its final stages. I am especially grateful to the departmental assistant Angela Joy Schippers who patiently helped navigate me through my move to UW. Special thanks to Eliakim Sibanda, Royden Loewen, Ryan Eyford, Janis Thiessen, and James Hanley. Thanks also to Carlos Colorado, Jason Hannan, and Peter Ives for helping foster a vibrant and vital academic community at UW for young and engaged scholars. Thanks to all my students who helped shape this project in unexpected ways. Special thanks to Richard Raber, Steven Dueck, and Joel Trono Doerksen. Thank you to Alexandra Judge for her research assistance.

Finally, thanks to Dr. Matt Dyce, a brilliant and generous scholar whom I am proud to call a friend. And to my old friend Jamie Phibbs and his lovely family who have been a well of generosity and friendship since our fortuitous reunion in Winnipeg.

This project would simply not have been possible without the unwavering love and encouragement of my family. My incredible in-laws—Tony and Pina Moretti—fed, clothed, housed, and loved us, while providing invaluable childcare. Though they will likely never read this book, it exists because of their remarkable lifetimes of sacrifices—a debt which I can never begin to repay. Thanks to Domenico Romanelli in Italy for all his support for our family. All my love to my brother Matthew Lawrie, a decent, kind, and talented man, and to his wonderful partner Sarah McMulkin. To my father, Dr. Bruce Lawrie, who through a lifetime of unconditional love and support introduced me to the joys of the life of the mind and the need to reconcile it with an insistent humanity. To my mother, Linda Lawrie—the strongest person I know—who fortified me with her love and strength. I love you both very much.

To my wife, Rose, my love, my partner in parenting, my confidant, and my best friend. Without her constant support, sacrifice, and encouragement, this book, and indeed my entire career, would simply not be possible. I am sorry for the mistakes I have made in the past, and the many I am sure to make in the future, but know that I love you. From the day you first sat beside me in a dreary undergraduate European history lecture at York University, you have brought laughter, passion, and unalloyed happiness into my life. *Ti amo mia tascabile venere.*

Finally, this book is dedicated to our children: To our precious son, William, born during its initial stages, and whose finger prints are quite literally all over the project. His insatiable enthusiasms and diatribes about everything and anything from the respective merits of dinosaurs versus mammals, Star Wars heroes versus villains, Vikings versus pirates, the intricacies of the solar system, Greek Myths, Minecraft, comedic timing, and the perfect knuckleball were always a welcome (often in hindsight . . .) distraction. To our darling daughter, Zoë—my sweet little peanut—born during the book's final stages, and whose mischievous zest for life, unconditional empathy, and boundless curiosity are a source of constant delight. William and Zoë, your presence in my life is a sublime gift; you are my pride, my joy, and my hope.

Paul R. D. Lawrie
Winnipeg, Manitoba
October, 2015

Introduction

Imagining Negro Laboring Types in Fin de Siècle America

"He (the Negro) is here; we can't get rid of him; it is all our fault;
he does not suit us as he is; what can we do to improve him?"
—Charlotte Perkins Gilman, "A Suggestion on the Negro
Problem" (1908)

"How does it feel to be a problem?"
—W. E. B. DuBois, *The Souls of Black Folk* (1903)

The spring of 1885 had been an especially trying one for Ben Bailey. A fledgling "mulatto pugilist," he was struggling to eke out a living in the fetid, smoke-filled, beer-doused taverns and athletic clubs of Philadelphia.[1] In March, he had fought a "rattling four round bout at Chuck's Club Theater" against a "stalwart mulatto," Amos Scott. According to the *Philadelphia Record*, Scott "had had the best of the fight after the second round and was declared the winner." A little over a month later, Bailey successfully battled one "Clipper" Donahue for six rounds in uptown Philadelphia "with soft gloves until the cry of 'police' went up stopping the match" and the referee declared it a draw.[2] For the remainder of the spring and into summer, Bailey's fighting opportunities dried up. Occupational opportunities for working-class black men in the late nineteenth-century urban north were limited. Boxing—despite its brutality, meager purses, and illegality—provided a rare chance for working men across the color line to acquire a modicum of financial autonomy outside traditional labor markets.[3] Perhaps it was his failings in the ring, general lack of job prospects, or a chance to parlay his one source of capital—his body—to his benefit that led Bailey to answer a request for models for a curious photographic study on human and animal movement at the University of Pennsylvania.

The man behind this unusual study was Eadweard Muybridge, an eccentric English polymath who had gained a measure of fame as one of the foremost landscape photographers of the American West, and his pioneering motion studies *The Attitudes of Animals in Motion* featuring the iconic imagery of a trotting horse. Through a mixture of professional ambition and institutional imperatives, Muybridge arrived at the University of Pennsylvania in the summer of 1883 to undertake one of the first sponsored research projects of the American university.[4] Officials at Penn afforded Muybridge and his assistants all manners of equipment, facilities, and a full studio on university grounds, where he assembled a motley assortment of subjects—student athletes, art models, soldiers, acrobats, fencers, and boxers—to populate his ambitious human motion studies. To capture an array of physical movements, Muybridge photographed each subject simultaneously from three different positions using a battery of twelve cameras for the lateral view and another two cameras with twelve lenses for the front and rear views. From 1884 to 1887, Muybridge and his assistants produced over 100,000 images and published 781 collotype plates containing more than 20,000 figures of moving men, women, children, animals, and birds.[5] The epigraph for the original leather-bound *Animal Locomotion: An Electro-Photographic Investigation of Consecutive Phases of Animal Movements* (1887) described it as a "work for the Art Connoisseur, the Scientist, the Artist and the Student of Nature," positing the body as both an aesthetic ideal and a corporeal index of civilizational progress (or lack thereof) amid rapidly shifting national and global labor economies.[6]

Ben Bailey spent little more than a day in Muybridge's studio, and yet he is one of *Animal Locomotion*'s most notable models. First and foremost, Bailey was the lone person of color among the ninety-five models to appear in any iteration of *Animal Locomotion*.[7] Muybridge photographed him performing a number of activities, from the pedestrian (walking up and down stairs) to the provocative (throwing a punch and hurling a stone). Bailey's greatest significance, however, lay in the manner in which these activities were presented, for it was in the series of photos numbered 524 to 531, featuring Bailey, that the anthropometrical grid—the statistical mapping of the human form—made its first appearance in American photography.[8] The grid emerged from the mid-nineteenth-century nexus of photography, ethnography, physical anthropology, and colonialism as a powerful tool of racial-knowledge production used to demarcate the contours of the "civilized," white, Western body versus the ostensibly savage body of color. But whereas previous versions of the grid had positioned the body against a backdrop divided into two-inch squares by means of silk threads, Muybridge utilized a grid composed of threads dividing the field into five-centimeter squares to chart a variety of physical states and movements. For Muybridge, motion—as representative of the body's productive capacity—was the primary field of anthropometric inquiry. Following the grid's introduction with Bailey in negative number 524, Muybridge retained the grid for all subsequent series in *Animal Locomotion*, presaging a shift in popular understandings of the body—specifically the laboring male, black body—as a site of scientific inquiry, efficiency, and progress.[9]

Black bodies have long been a source of fascination and fear throughout American history. Though of course which bodies were deemed to be "black" has shifted throughout American history. Yet from the beginning this fascination has been deeply informed by blacks'—or those defined as black—unique status as embodied capital to be bought and sold on the open market. In this sense blackness, notwithstanding its various historical permutations, has always been an expression of capital or lack thereof. Since the arrival of the first "20 odd Negars [*sic*]" in Virginia in 1620 to the slave blocks of the antebellum South, through the brutalities of post-Reconstruction penal labor, through the nation's various imperial forays, to the factory floors of the early twentieth century, black bodies have been poked, prodded, and violated to determine, discipline, or enhance their worth as commodities.[10] By the late nineteenth century, the shifting socioeconomic status of southern blacks engendered a severe backlash among many whites against the supposedly deviant, hyper-sexualized and brutish black, male body. The surge in anti-black violence throughout the South was a visceral yet calculated attempt to put the increasingly recalcitrant black body back in its proper "place," or destroy it in the process—often at the end of a noose.[11] It was during this period that the grid—a key technology of nineteenth-century imperialism for making sense of the colonial "other"—entered the American imagination, mapping one of the era's most feared, misunderstood, and ultimately elusive figures: the black male body.[12]

Despite the relative physical absence of blacks north of the Mason-Dixon Line, their figurative presence in racial theory as representatives of a savage vitality seemingly lacking in the nation's elites prevailed. Fears that modern civilization had become a victim of its own success—its elite architects enfeebled by the debilitating effects of "over-civilization" evinced in the psychopathologic disease of neurasthenia—led to a renewed fascination with the ostensibly redemptive primitivism of the non-white races.[13] Bailey's taut classical musculature and tawny hue stood in stark contrast to the pallid concave chests and rounded shoulders of Muybridge's white student athlete models. These jarring physical juxtapositions—exacerbated by the scandalous nakedness of the models—laid bare prevailing societal anxieties regarding the degeneration of elite white men in the face of an atavistic savagery personified by men of color such as Ben Bailey.[14] Although Muybridge himself expressed little interest in theories of racial evolution, Penn sponsored his work as part of a larger campaign to document best practices in hygiene, physical education, and racial (read: white) progress—a counterweight to prevailing fin de siècle fears of degeneration and decay routinely expressed in metaphors of national and racial degeneration.[15] Whether as objects of fear or fascination, black bodies such as Bailey's would prove instrumental both as commodities and concepts in the development of Progressive Era labor economies and the industrial state.

This book tells the story of an idea—it is both the story of the black worker in the Progressive imagination, and the story of how a collection of thinkers across the natural and social sciences considered the role of the black worker in the nation's industrial past, present, and future, and how these bodies of thought proved crucial in the making of the U.S. industrial state in peace and war.[16] Whereas twenty-first-century thinkers tend to view race as a social construct, or "social fiction," their Progressive Era predecessors viewed race as a physiological and historical "fact" and a key agent of human progress. For W. E. B. DuBois, "the history of the world is the history, not of individuals but of groups, not of nations, but of races, and he who ignores or seeks to override the race idea in human history ignores and overrides the central thought of all history."[17] Consequently, racial function—occupational or otherwise—was thought to shape racial forms: a negro *was* as a negro *did*. But as the nation transitioned to unprecedented forms of mass industrialism and immigration, an emergent management class drawn from the social, natural, and medical sciences tried to reconcile social problems along biological lines via what historian Dan Bender terms a theory of *industrial evolution*, linking certain peoples to certain kinds of work. Theories of industrial evolution ultimately reconfigured the laboring body as an index of racial progress or lack thereof. New taxonomies of labor fitness inverted prior models of racial function, shaping ones in which form dictated function: a negro *did* as a negro *was*, inexorably linking labor fitness to color and the body. Paired with the narratological structures of the emergent theories of Taylorism, the story of races—told in strident cadences of masculinity and manhood—became one of development and decline, efficiency and inefficiency, and the constant tension between civilization and savagery.[18]

The Negro working "type" of the Progressive imagination was born in the era's photography studios, corporate boardrooms, universities, factory floors, draft boards, battlefields, and hospitals. Yet this was a deeply ideological exercise, given the historical role of race in American capitalism. As Nell Irvin Painter notes, "Race is an idea, not a fact, and its questions demand answers from the conceptual rather than the factual realm."[19] Positing ideology as a system of beliefs characteristic of a particular class or group and "the social processes by which meanings and ideas are produced," the following examines how the "minds [of the era's managerial elites] met the world" via the management of labor across racial lines. The focus here is not strictly on identities, individuals, or institutions but on the processes or arcs of ideological transformations that were, at specific historical moments, central to the formation of categories and hierarchies of race, labor, and the state[20]—a focus on ever-shifting discursive practices and strategies, commonly understood as "expertise," used by elite individuals and institutions seeking to make sense of the world as it was and how they wished it to be.

Progressive Era racial thought was defined by common questions rather than a singular set of answers. Questions such as: Did races exist? And if so, were they the result of heredity or environment? How did one measure or quantify these differences? And, perhaps most importantly, what was the value of race in the calculus of American industrial capitalism? To answer these questions, progressives developed a bewilderingly complex and contradictory body of thought whose obsessive focus on racial hierarchy and white supremacy seem abhorrent to most present-day observers. Yet Nancy Stepans reminds us that "the scientists who gave scientific racism its credibility and respectability were often first rate scientists struggling to understand what appeared to them to be deeply puzzling problems of biology and human society."[21] Dismissing these forms of scientific racism as mere "pseudo-science," and their purveyors as cranks or bigots, obscures the natural and social sciences' fundamentally historical character as demands of their respective labor economies. Ben Bailey's experiences in the ring and the studio were representative of broader shifts in the ongoing devaluation of black laboring bodies with each passing year post-emancipation.[22]

Forging a Laboring Race posits the black working body as a site of contested knowledge to re-examine how black proletarianization was mediated through the state and how progressives came to understand these processes in deeply corporeal terms. Blacks' experience of proletarianization—the means through which black labor shaped, and was shaped by, American industrial modernity—was fraught with contradictions. For many workers across the color line, the shift from a rural to an urban economy entailed the gradual deskilling of labor along with a rapid erosion of social networks and personal autonomy.[23] Whereas some black workers certainly did experience these tensions in varying degrees, entry into industrial modernity afforded most blacks a modicum of socioeconomic autonomy given their previously extreme marginalization. Entrapped in the virtual peonage of sharecropping and penal labor in the South—and constrained by the de facto or de jure polices and practices of Jim Crow nationwide, the sphere of opportunity for black labor could only be increased.[24]

Heretofore black proletarianization has generally been understood as a function of the early twentieth-century Great Migration of blacks out of the South to points north.[25] Yet migration narratives that detail how the peonage of sharecropping, the insidious boll weevil, and the often-fatal indignities of Jim Crow pushed blacks out of their ancestral southern homeland in response to the lure of greater social freedom in the North tend to elide the modernist character of the New South, while positing the war itself as little more than a midwife to these migratory processes.[26] However, the managerial cultures of the wartime state, linking war with work from the trenches to the factory floor, were national in scope. For the millions of blacks who did not migrate north, the wartime state—and military service in particular—provided their defining exposure to the systems of industrial modernity. The war transformed the private industry's relatively inchoate response to blacks' entry into industrial modernity into a coherent attempt to rationalize the masses of black labor.[27] Officials at the Department of Negro Economics (DNE), the Committee on Anthropology (COA), and the Federal Board of Vocational Education (FBVE) employed sociological, anthropometric, and rehabilitative methodologies to define and regulate black labor across regional lines—from the cotton fields of Dixie through the factories of the North, to the trenches of France. The wartime state's management of black labor enhanced the authority of the state as an arbiter of expertise and engendered new sweeping cultures of social control based on the definition and discipline of working bodies of color.[28]

The metaphor of the black working body pervades the histories of black proletarianization, migration, and war. Yet its meaning and function as a signifier of social differences is unclear. Building on recent works on racial bodies as conduits of social knowledge and control, I seek to delineate the "absent presence" of black working bodies in the history of black proletarianization.[29] Wartime demands further entrenched and naturalized the linkages between racial form and function. Black soldiers and workers continually contested the intrusive and coercive nature of these arrangements through small and large scale acts of resistance—from "malingering" in the ranks to struggles over veterans' care. Fundamental to this study is the delineation of how working bodies—particularly black working bodies—have served as sites of both repression and resistance within modern American industrial capitalism. Further, it explores how some people, under certain circumstances, could draw on their bodies as a resource, while others could not, because they were positioned as captive to their bodies.[30] The abstract idea of the normative worker—like the abstract normative citizen of modern liberal theory—casts gendered, racialized, and disabled bodies as exceptional, and therefore deviant. Managerial elites marked black workers and soldiers as diseased, dirty, unskilled, and hypersexual, reconfiguring blackness itself as a disability. Whereas physiological whiteness conferred normalcy and access to the social and financial benefits of martial and industrial labor, those bodies bearing the stain of blackness found their access to be restricted and their citizenship undermined.[31]

Muybridge's images of Bailey endure in part because they evince familiar modern practices and aesthetics of quantification, qualification, and bodily discipline. The precise symmetry of the grid evokes a medical and scientific authority instantly recognizable to the modern eye—an authority that drained images of their respective social, economic, and cultural contexts, reducing them to mere objects of analysis that increasingly fell under the auspices of the state. Specifically, the power of the state to shape the corporeal contours and health of the body politic—what Michael Foucault defined as the policies and practices of "bio-politics"—informed the growth of the modern nation-state.[32] However, the figure of Bailey also represented something of a contradiction, given that these modern modes of inquiry were applied to a *black* body—one that was long understood to be the embodiment of savagery, and the very antithesis of modernity itself. That is, a modernity that existed only in the purview of those individuals, groups, and modes of thinking defined as white. Conversely, non-white peoples were deemed an impediment—or at best simply peripheral—to the workings of industrial modernity, dusky specters in the machine of American industrial civilization, immune or incidental to its imperatives of mass standardization.[33] The mere presence of a body of color in this context—in which quantification normally denoted a narrative of progress—presaged how vital the making of race and racial difference would prove to be in the development of American industrial modernity. Following his *Animal Locomotion* sessions, Ben Bailey fought a smattering of bouts with mixed results, in and around Philadelphia, before vanishing entirely from the historical record. Yet his striking images captured via Muybridge's lens lived on—auguring new ways of seeing, making, and thinking about race, labor, and the body in industrial America.[34]

This book begins with an examination of late nineteenth-century actuarial science, which characterized Negroes as a depraved race destined for extinction. Many observers felt that, once freed from what they considered the protective embrace of slavery, blacks would be unable to compete in the unforgiving struggle of industrial evolution, much like the American Indian before him.[35] The theory of black extinction found its greatest advocate in the work of Frederick L. Hoffman, an actuary at Prudential Life Insurance. In *Race Traits and*

Tendencies of the American Negro (1896), Hoffman charted black racial decline through the indices of venereal disease, criminality, and vital capacity—a shorthand for respiratory health. Theodore Porter describes turn-of-the-century statistics as "the science of numbers applied to the life history of races and nations." Statistics, specifically actuarial science, allowed Hoffman to quantify the relative rate, nature, and cost of deteriorating black bodies in concrete scientific and monetary terms.[36] Black thinkers stridently condemned Hoffman, but their critiques were hampered by an inability to transcend prevailing statistical methodologies viewed as unerringly factual. Hoffman's actuarial narratives of black extinction would dominate much of early twentieth-century racial discourse, thanks in part to their seemingly empirical nature, which, as a mark of the defining ideology of the age, linked inquiry to utility—in a process known as expertise. Policies and practices of reform rooted in larger questions of authority, expertise, and power—such as those posed by Hoffman—were ultimately linked to the growth of the bureaucracy of the wartime state.[37]

World War I turned the African American worker into a national industrial factor. Industrialization, migration, and war required new forms of productivism and taxonomies of racial fitness to reconcile blacks increasing—albeit limited—occupational mobility to their traditional social subjugation.[38] The war offered social scientists a grand experiment in the possibilities of the war-collectivized state. Walter Lippman, writing in the *New Republic*, declared that "we stand at the threshold of a collectivism which is greater than any as yet planned by a socialist party."[39] Progressives hoped that war, with its various forms of state-sanctioned coercion, would help solve various social ills, such as the "Negro problem." As Tami Davis Biddle notes, "Wartime affords the historian a window on a society under stress, actively enacting, reconsidering and perhaps redefining its sense of self."[40] The cultures of management born in the late nineteenth century that reached maturity during the war were informed by an abiding desire to reconcile form to function in the pursuit of social order and efficiency. Scholars have noted that Taylorism, or industrial standardization, was an extended narrative structure and discourse that extended far beyond the factory floor to encompass every aspect of cultural existence.[41] Progressives employed the language and theory of productivism, a totalizing ideology that sought to subordinate all social—especially racial—relations to production, linking the human project of labor and even the nation state itself to a universal attribute of nature.[42] American progressives drew heavily on European theories regarding the productive duality of the modern worker-soldier and conscription as a form of social and labor control.[43] Black social scientists worked to penetrate the wartime state in an advisory capacity to compensate for their acute lack of institutional support and financial capital. The hope and challenge of all progressives was to capture and democratize the machinery of war collectivism to shape public policy.[44]

Total war militarized industry at the same time that war was becoming increasingly industrialized. Chapter 2 examines the wartime state as a key mediator in the processes of blacks' entry into industrial modernity. The establishment and workings of the Department of Negro Economics (DNE)—the first federal agency devoted exclusively to black labor since Reconstruction—was an attempt to reconcile the "Negro problem" with the "Labor problem" within the institutional nexus of early twentieth-century sociology. DNE officials—many of whom were black—drew on Chicago School sociology to chart and acclimatize black migrants' shift from rural to modern industrial life. DNE officials sought to incorporate black workers into the wartime labor economy through the development of black labor expertise, instilling migrants with a new industrial consciousness through worker efficiency campaigns and establishing stronger links between white capital and black labor. Scholars have noted that "the predicament of black workers in industrializing America continuously engaged the attention of black social scientists, more than any other single topic"—offering further proof of the interconnected "Labor" and "Negro" problems. DNE officials fused vocational uplift with sociological expertise to endow black workers with a respectable and efficient work ethic, a new "industrial consciousness" to counter prevailing narratives of the congenitally unfit Negro worker and expedite black labor's entry into mainstream labor economies.[45]

Wartime imperatives required new taxonomies of racial labor fitness. Chapter 3 examines how social scientists on the Committee on Anthropology (COA), a part of the National Research Council (NRC), used the science of anthropometry to evaluate the health, shape, and fitness of "the Negro type" through the wartime draft. COA members defined anthropometry as "that part of anthropology in which are studied variations in the human body and all its parts, and particularly the differences of such variations in the races, tribes, families and other well defined groups of humanity." Framing war as work, the COA defined racial types through the measurement and evaluation of the first million army recruits, the "multiracial" workforce of the American International Shipbuilding Association at Hog Island in Philadelphia, and the measurement of roughly 100,000 demobilized men in the summer of 1919.[46] Wartime anthropometry worked to equate the military evaluation of racial bodies with their industrial classification positing the black worker/soldier— much like Ben Bailey just over a generation prior—as both object and agent of racial inquiry.[47] Indeed the management of difference—the division of labor along racial lines in both the industrial and military spheres—was constitutive to the making of the wartime and postwar state.[48]

Chapter 4 details how federal efforts to rehabilitate disabled African American veterans led to new forms of racial knowledge and racial labor control in postwar America. A key agent in this process was the Federal Board of Vocational Education (FBVE), charged with rehabilitating the citizen-solider into the citizen-worker. FBVE policy linked scientific management and eugenics to develop a catalogue of racial taxonomies in which labor fitness was linked to color and the body. Through the stages of diagnosis/benefits, training, job placement, and hospitalization, officials struggled to determine whether they could, or even should, mend broken black veterans, who were generally understood to be defective by definition. Indeed, wounded black veterans were often accused of trying to "unjustly profit from their innate inferiority." African American veterans rejected these characterizations, arguing for their right to rehabilitation as soldiers, citizens, workers, and men.[49] The FBVE's efforts to salvage black veterans for work in the postwar industrial army was a key attempt by progressives to frame social policy along biological lines, rationalize racial labor hierarchies by conflating disability with blackness, and infuse a corporeal element to the dictum that war was indeed the health of the state.

World War I produced a wealth of social scientific racial knowledge. Chapter 5 analyzes how social scientists at the National Research Council (NRC) built on wartime evaluations of African Americans to develop, sustain, and institutionalize a body of knowledge on the Negro worker under federal auspices. Various NRC committees, such as the *Committee on Scientific Problems of Human Migration (1922)* and the *Committee on the American Negro (1928)*, were established as clearinghouses for racial labor expertise. The war changed perceptions of African American labor fitness in three significant ways: wartime migration and military service established the Negro as an industrial factor; rapid migration and urbanization led to mental and physical changes on African Americans; and national efficiency required racial integrity—hence the need to avoid the supposedly deleterious practice of race mixing both at home and abroad. Postwar social scientists' fascination with the processes of race mixing evinced a distinct inability on their part to delineate the exact physiological contours of the New Negro, engendering anxieties about the very stability of blackness as a racial identity.[50] In the end, wartime and postwar managerial elites unwittingly severed race from biology, creating new culturist models of racial and labor difference while continuing to characterize blackness in decidedly pathological terms. Though the terms of black deviancy may have changed, the figure of the Negro as an object of inquiry, as a problem to be solved, persisted in different iterations, and continues in present-day debates regarding race in America.

The successful management of the war effort provided social scientists with the impetus to rationalize humans and human networks through theoretical models analogous to those in the natural sciences.[51] That is, it allowed elites to effectively create an organic or corporal model, or map, of the racial dynamics of the nation's labor economy.[52] Heretofore white supremacy had been defined by its power to define the racial "other" as a mere object of inquiry, but the postwar state now afforded unprecedented methodological, institutional, and financial means to create taxonomies of racial labor fitness. At first glance, the immediate postwar era can be seen as the high tide of white supremacy evinced in sweeping racialist legislative polices such as the 1924 Immigration Restriction Act, Virginia's Racial Integrity Act, and best-selling paeans to the virtues of whiteness, such as Madison Grant's *The Passing of the Great Race* (1916) and Lothrop Stoddard's *The Rising Tide of Color Against White World-Supremacy* (1920).[53] Taking stock of this surge in white supremacist culture, DuBois noted that "all through the world this gospel is preaching. It has its literature, it has its priests, it has its secret propaganda and above all—it pays!"[54] But postwar white supremacy was a profoundly pessimistic ideology. The often-ambiguous results of physical and mental wartime testing—revealed in the ostensibly mentally deficient but "physiologically superior Negro"—revealed deep fissures in the edifice of postwar American white supremacy. It also hinted that perhaps heredity could not be reconciled to function, and biology did not equal destiny.[55]

"How does it feel to be a problem?" asked W. E. B. DuBois in *The Souls of Black Folk*, his eloquent meditation on the racial politics of turn-of-the-century America.[56] Yet at the heart of the "Negro problem" was the problem facing all free labor in hierarchal societies. For workers, free labor meant economic autonomy, including the freedom not to work; for employers, free labor meant economic dependence and the need to define which workers could or could not perform which kinds of work.[57] The "Negro" and "Labor" problems were social fictions rooted in the sociocultural demands of contemporary labor economy. Progressives' ardent faith in empirical reform was tempered by an acute sense of pessimism regarding the perils of degeneration embodied in degenerate peoples such as Negroes.[58]

By the turn of the century, emergent theories of industrial evolution, shifting labor demographics, and the transition from proprietary to corporate capitalism conspired to negotiate these tensions in decidedly black and white corporeal terms.[59] Though these processes of labor control along racial lines extended beyond the continental U.S. and were deeply informed by turn-of-the-century American imperial engagements, our focus here is on the linkages between the domestic sphere and wartime testing of the First World War when the line between war and work became irrevocably blurred.[60] Wartime testing provided a scientific rationale to these processes and the often-horrific extralegal racial violence that undergirded the era's labor economy. High and low cultural policies and practices of racial control conspired to devalue black bodies and black humanity at every turn. And while managerial elites often remained unsure over the exact dimensions of the white working type, they remained convinced of its fundamental non-blackness. As Nell Irvin Painter notes, statutory biological or occupational definitions of whiteness were "notoriously vague—the leavings of what is not black." From the crooked contours of the ostensibly degraded black working body—with its "flat feet," "tropical lung," and "syphilitic carriage"—emerged the normative (read: white) outlines of the ideal worker of the Progressive imagination.[61]

Muybridge's images of Ben Bailey, the "vanishing Negro" of actuarial narratives, the wartime black laborer, the Negro of the wartime draft, the broken black bodies in need of rehabilitation, and the New Negro of the postwar social scientific imagination—all were echoes of the many afterlives of slavery, the real and imagined twists and turns of blacks' fractious transition from bondage to wage labor in post-Reconstruction America.[62] Though they varied in method and degree, these afterlives linked the discipline and surveillance of the antebellum plantation to modern forms of quantification and commodification to fix the Negro into his or her proper place. Viewed through a corporeal lens, these various afterlives take on the character of a "changing same": a bodily frame upon which historically specific meanings are hung and rehung to fit the right races to the right places.[63] Foregrounding the working black body as a metaphor for making sense of black laborers' transition into industrial modernity—and how these imaginings informed the development of the modern industrial state—requires us to investigate the studios, factories, universities, trenches, draft boards, cantonments, and hospitals of Progressive Era America.

1

Mortality as the Life Story of a People

Frederick L. Hoffman and Actuarial Narratives of African American Extinction, 1896–1915

"A race may be interesting, gentle and hospitable; but if it is not a useful race in the common acceptation of that term, it is only a question of time when a downward course must take place."
—Frederick L. Hoffman, *Race Traits and Tendencies of the American Negro*, 1896

In November 1884, Frederick Ludwig Hoffman, a nineteen-year-old jour-neyman laborer from Oldenburg in northwestern Germany, arrived in New York City with little more than the proverbial dollar in his pocket. Despite lacking an extensive formal education, the young émigré was a proud autodi-dact determined to make a name for himself in America. Following a string of unsuccessful jobs—including a brief stint collecting insurance premi-ums door-to-door in Boston—Hoffman decided to try his luck south of the Mason-Dixon Line. Hoffman found Dixie to be a "veritable paradise per-fumed with the sweet smells of magnolia and oleander" and accompanied by leisurely "paddles down Edenic rivers."[1] This southern reprieve was soon shattered during a trip down the Mississippi aboard the *City of New Orleans*, where he witnessed "the truly horrible brutality practiced on the Negro deck hands" by their white shipmates. Despite his initial sympathy for the "much maligned Negroes," Hoffman was transfixed by the spectacle of laboring black bodies: "Though they were grossly ignorant, they would carry for long hours heavy sacks of corn, up and down the gangplank, leading at a steep angle to the shore."[2] Hoffman's fascination with the deck hands' "incredible perfor-mance of labor" and their "savage vitality" echoed prevailing notions of blacks as little more than atavistic beasts of burden—pitiable reminders of a rapidly vanishing preindustrial culture.[3]

Hoffman's experience aboard the *City of New Orleans* proved fortuitous. His introduction to the brutal complexities of American race relations led to a lifelong interest in African Americans, provided the subtext for his future actuarial investigations of racial health, and gave him access to the era's institutional and ideological networks of vocational education. Hoffman chose to remain in the South and soon found work at the Life Insurance Company of Virginia in Norfolk. In 1890, while traveling aboard a train bound for Hampton, Hoffman was overheard discussing the incident aboard the *City of New Orleans* by Frances Morgan Armstrong, the spouse of General Samuel Armstrong, founder of Hampton Institute. Born and raised in Hawaii by missionary parents, Samuel Armstrong claimed "a deep familiarity with the tropical colored races" and had founded Hampton in 1868 to inculcate the freed-people with much needed "habits of industry and health."[4] Yet as the nineteenth century drew to a close, even self-professed "friends of the Negro" such as Armstrong had begun to doubt the efficacy of vocational education, given blacks' "congenital aversion towards labor." Like the Indian before him, the Negro seemed destined for racial death as a prerequisite for the nation's continued progress.[5]

This chapter examines the efforts of Frederick L. Hoffman—a statistician and actuary with Prudential Life Insurance—to chart the initial stages of African American proletarianization through actuarial narratives of race suicide. Actuarial science, a key tool of racial labor division in turn-of-the-century American political economy, reveals the ways in which the era's managerial elites came to understand shifts in the national labor market in explicitly corporeal terms—to try and map shifts in the labor economy onto working bodies. In works such as "Vital Statistics of the Negro" (1892), the seminal *Race Traits and Tendencies of the American Negro* (1896), and various other studies undertaken on behalf of Prudential, Hoffman used the metrics of crime, miscegenation, and the broadly defined "vital capacity" (a measure of respiratory health) to quantify the respective social, physical and economic effects of African Americans' transition from bonded to contract labor.[6] Actuarial assessments of African Americans as insurable "risks" were both economic *and* cultural evaluations of their past, present, and future labor fitness. Indeed, as Brian Glenn notes, "Actuaries rate risks in many different ways, depending on the stories they tell about which characteristics are important, and which are not." These "stories" invariably adhere to prevailing gender, class, and racial hierarchies that "may be irrelevant to predicting actual losses." For Glenn, "almost every aspect of the insurance industry is predicated on stories first, then numbers."[7] The story of Negro progress—or lack thereof—was one of a people in decline. Freed from the supposed protection of slavery, the congenitally criminal, mongrel, and tubercular Negro seemed destined to expire amid the mechanized rhythms of modern industrial civilization according to this view. The imperatives of rapid industrialization, imperialism, mass consumerism, and white supremacy necessitated the creation of the "vanishing Negro": a debased inferior "other" against which the progress of white civilization could be measured and monetized.

The cultural and intellectual landscape of turn of the twentieth century America was littered with an array of self-described "race experts" working to solve the "Negro question." An eclectic mix of anthropologists, sociologists, scientists, physicians, writers, and actuaries vied to define, delineate, and quantify the vast socioeconomic and cultural changes brought on by the transition of blacks from slavery to freedom. Frederick Hoffman's status as a young actuary with an interest in racial reform gave him a unique prospective on the character and constitution of the color line. Following their chance encounter on the rails, Hoffman and the Armstrongs became close friends—and, under the latter's tutelage, Hoffman gained an appreciation of the domestic "Negro problem" as part of a larger global race problem that stretched from the pineapple plantations of Hawaii to the cotton fields of Dixie.[8] The question of what to do with the Negro was tied to the broader demands of contemporary political economy: delineating which peoples could do which kinds of work via the logic of industrial evolution whereby the working body was reconstituted as an index of civilizational fitness or lack thereof.[9]

Applied sciences such as anthropology, actuarial science, and anthropometry (the statistical measurement of the human body) emerged in response to cultural anxieties regarding the social, economic, and physical costs of industrial civilization. Economists and social scientists such as Richard Ely and Thorstein Veblen developed new evolutionary models of political economy that sought to reconcile racial form to economic function. Attendant theories of race suicide represented the most extreme versions of these new forms of thinking—a zero sum logic in which one race would have to perish for another to thrive. For Hoffman, "statistics was the science of numbers applied to the life history of communities, nations, or races."[10] Emerging ideas of the nation-state as an organic entity led many to conclude that black extinction, while necessary and perhaps even desirable, was nevertheless fraught with social, economic, and health risks for blacks and whites alike.

Progressive Era America experienced a shift in understandings of political economy as the social relationships between production and consumption, to one dominated by economic theories seeking to reframe these relations on a more axiomatic and mathematical basis. Scholars have described this as a shift from proprietary to corporate capitalism, citing the turn of the century transatlantic world as a period awash in "an avalanche of numbers." Quantification was privileged as a superior form of producing knowledge about the social and natural worlds.[11] Theodore Porter notes how statistics combined a utopian urge for order with the modern drive for efficiency. Much like banking and accounting, Porter posits insurance as the "classic information industry."[12] However, as Joan Scott reminds us, the very practice of quantification is rooted in the imperatives of power, surveillance and discipline.[13] Statistics—or "state numbers"—consolidated the development of the modern industrial state through indices of population, trade, manufacture, and mortality that aided in the shaping of public policy.

Yet from their inception statistics were deeply informed by racial considerations. In the late 1860s, Francis Galton—a cousin of Charles Darwin—used the normal, or Gaussian, method of distribution to transform statistics from a science devoted to the accumulation of socially useful data to one characterized by systematic mathematical theory. He was especially interested in the distribution of deviations, or regressions, from the mean as they related to human populations and the regulation of the gene pool. Galton named this applied science, *eugenics*, a mixture of Greek terms literally defined as "well born" or the practice of "better breeding." From its inception, eugenics linked state- and nation-building to biology, which by the early twentieth century had taken on a far more explicit racial tone.[14] Galton's protégé Karl Pearson, the founder of modern mathematical statistics embraced a fully positivist ethos whereby the world consisted not of material objects but of perceptions. The role of science—or statistics more specifically- was to provide shape to the natural and social world material: method not material would prove to be the unifying logic of modern scientific inquiry. This subjective taxonomic impulse to order the world not as it was but as it should be was imbricated in the late nineteenth-century processes of race and state making.[15]

In the waning years of the nineteenth century, managerial elites developed intricate taxonomies of racial types to order the era's shifting labor demographics. These new "bodies of knowledge" married form to function, linking the right peoples to the right kinds of work. And though the manner and degree of inquiry may have differed, antebellum and industrial era black bodies were united through their common commodification. Valuation was implicit in the analyses and quantification of working bodies from the slave block to the factory floor.[16] Unlike whites, black laborers claimed the dubious distinction of having been *literal* commodities only a generation prior. Post emancipation blacks had found themselves condemned to the debt peonage of sharecropping shut out of most industrial service jobs and consigned to the margins of the nation's labor economies. Hoffman's actuarial narratives of black race suicide—through their transposing of flesh and blood bodies into abstract statistical values—were key examples of the contentious nature of race and nation making in turn-of-the-century America. For example, under the direction of Francis Amasa Walker, the late nineteenth-century census became a key mechanism of racial categorization in the newly reconstituted American republic. At a time of increasing black occupational mobility, actuarial models of black extinction were a means to preserve the race's social containment.[17]

Across the nation the rise of mass consumer culture and its attendant technologies—such as the railroads—had eroded many of the traditional markers of racial identity. Within the South, a growing demand for unskilled and semiskilled labor in urban centers had allowed blacks to make small but persistent gains in real estate and small business capital. To the dismay of many whites, the New Negro was wealthier, more mobile, and more assertive than his predecessor of only a few decades ago.[18] The figure of "the vanishing Negro" became both an explanatory model of contemporary political economy and a utopian vision of a nation free from the scourge of blackness. "Fantasies of auto genocide or racial suicide," writes Patrick Brantlinger, "are extreme versions of blaming the victim, which throughout the last three centuries have helped to rationalize the genocidal aspects of European conquest and colonization." Theories of black race suicide worked to alleviate lingering fears over societal and industrial degeneration "by replacing the imperiled white race as victim, with the self-extinguishing savage as sacrifice" to the greater social good.[19] By the late nineteenth century, valuation of the greater social good was almost entirely measured within the cash nexus of a new corporatist national economy—one in which the very experiences of life and death were transposed into marketable insurable commodities on the open market.

The American insurance industry originated in the working classes. Beginning in the late nineteenth century, American workers banded together in their respective occupations to mutualize the various risks endemic to wage labor. Heretofore, working people lived in constant fear of an unexpected illness, accident, or job loss. In response, they formed mutual benefit societies to provide members with funds for services such as burial insurance. Dan Rodgers notes that though these societies were notoriously unstable and actuarially unsound—often functioning as little more than a lottery—no form of organization set down deeper roots in the working classes.[20] Mutual aid societies were widespread among the coal miners of Appalachia, the industrial immigrant workers of Chicago, and African American workers nationwide. Regardless of their ethnic affiliations, these societies were defined by their informal and often corruptible nature. In *The Philadelphia Negro* (1899), the black sociologist W. E. B. DuBois noted that while some of the city's black societies were "honest efforts, most were swindling imitations of the pernicious white petty insurance societies."[21] Industrial insurance extracted the single most important aspect of the mutual aid society, burial costs, for use in a business contract with only one condition: payment. As the availability of insurance was reduced to a matter of revenue divorced from social standing, this produced a leveling effect in the "corporeal commodification" of working peoples and working bodies.[22] Jackson Lears argues that just as the movement for "sound money sought to tie ephemeral paper to the intrinsic value of gold. . . . modern forms of racism provided similar solidity to personal identity" in a secularizing uncertain market society. Actuarial science and the demands of a white supremacist political economy produced what Lears defines as a "biological personhood" along strictly racial lines.[23]

Following the Civil War, the major insurance companies insured African Americans on an equal basis with whites. This arrangement lasted until 1881 when Prudential became the first company to reduce life benefits to blacks by one-third while requiring that they still pay premiums in their original amount. Prudential was supported in its efforts by the Insurance Commissioner of Massachusetts, who concluded that this practice "was not a distinction on account of color, but on account of the differences in longevity between the two races apparently supported by mortality statistics."[24] The report dismissed any notion of racial prejudice, stating that, "to compel a company to insure for the same rates, different classes of people, with different prospects of longevity, would be to establish a grossly unjust discrimination against the longer-lived class in favor of the shorter lived class."[25] The editors of the trade journal *Spectator* concurred arguing, "the color line is not drawn simply because the applicants are negroes—the world is too progressive for that—but a distinction is made on account of the fact that companies cannot afford to grant policies at the same rates to colored as to white applicants and any legislation which is intended to force them to do so is particularly tyrannous."[26] In the end, industry officials saw the free market as an organic expression of evolutionary theory, delineating the "fit" from the "unfit" along monetary lines.

Blacks vigorously contested Prudential's exclusionary practices in the courts. They achieved a measure of success at the local level, by citing the Fourteenth Amendment's guarantee of "equal protection of the laws." In 1884, the Massachusetts legislature passed a law forbidding the custom of providing fewer benefits to blacks who were paying the same premiums as whites. Similar laws were passed in Connecticut in 1887, and Ohio in 1889, due in part by pressure from middle class blacks, and an overall push to regulate the industry at the federal level. When New York State introduced an anti-discriminatory bill in 1891, Leslie Ward, the vice president of Prudential, published a letter in an industry trade journal threatening to end sales to African

Americans altogether should the legislation pass. Ward's pleas were ignored and New York passed the measure in 1891, followed by Michigan in 1893.[27] Actuarial expertise was key to making the case for the insurance industry's refusal to insure blacks on an equal basis as whites. The work of one young actuary in particular, Frederick Hoffman, proved instrumental to proponents of anti-black insurance policies in New Jersey. Hoffman's "Vital Statistics of the Negro" (1892)—published in the *Arena*—cited extensive evidence of blacks declining health, "reversion to savagery," and inevitable extinction, all of which made them unprofitable insurance risks.[28] In 1893, Hoffman became a regular contributor to the trade journal *Spectator*, writing primarily on issues related to health and race. The following year, he took a job in Prudential's new statistical department at its Newark headquarters, where he began work on what would be his second work, *Race Traits and Tendencies of the American Negro*. Hoffman would remain with Prudential for nearly forty years writing on a variety of issues such as race, cancer, suicide, alcoholism, and national health insurance.[29]

Debates over African American Mortality in Turn-of-the-Century America

In the wake of Reconstruction, rapid industrialization, mass immigration to the north and legalized racial segregation in the New South, a national debate arose over whether the assumedly primitive Negro could survive the brave new world of American industrial modernity. This shift from social to biological interpretations of black inferiority was echoed by leading intellectuals such as Nathaniel Shaler, Dean of the Lawrence Scientific School at Harvard who noted, "Despite the strong spring of life within the race the inherited qualities of Negroes to a great degree unfit them to carry the burden of our own modern civilization."[30] The famed Progressive Era economist Charles Beard echoed prevailing sociocultural interpretations of labor economies when he declared that "modern civilization is industrial," thereby reconstituting industry as an index of civilizational (read: racial) development—or lack thereof.[31] Like many of his peers, Beard believed the fittest races to be laboring races: industrial labor being distinct and superior to the primitivism of handicraft labor, given its associations with both the pre modern and the feminine. Indeed the gendered, patriarchal, and racial nature of this discourse posited any and all working types as resolutely white and male.[32]

Progressives believed that like the American Indian before him, the Negro was marked for extinction. Yet in contrast to the noble Indian brave who had admirably resisted his fate with manly vigor, blacks were seemingly destined to simply expire through sloth and indifference. From the minstrel stage to the brutal spectacle of lynching, the degraded black body was seen as both cause and consequence of racial depravity.[33] In the waning years of the nineteenth century, biological theories of black degeneration were increasingly linked to the failures of Reconstruction and blacks' ill-fated participation in the political sphere. Popular opinion maintained that slavery had insulated blacks from the enfeebling effects of the free market and direct competition with the superior white wage laborer. Dr. Edwin Bushee, a Boston physician, argued that high black mortality and infectious disease rates were due in part "to the many subtle influences which uniformly cause an inferior race to quickly disappear when in direct contact with a higher civilization," thereby confirming the evolutionary basis of social and or civilizational development.[34]

Elites viewed the enfeebled bodies of the first generation of freed people as cautionary tales, testaments to Reconstruction's failure, and the hubris of reformers and philanthropists who had foolishly worked to uplift the Negro in defiance of the laws of biology. Moreover, many believed that slavery had compelled the seemingly indolent Negro to labor and that in its absence blacks would revert to their inherent slothfulness. Dr. Eugene R. Corson of Savannah noted that "many of these Negroes now passing away, are survivors of the old regime, where they were well cared for, and had reached at emancipation a safe age which kept them out of the struggle for life . . . the younger generations have been deprived from birth from such protection."[35] Historian James Bryce concurred: "The census just taken [in 1890] relieves . . . a source of anxiety. It is now clear that the Negro, long regarded as a factor in the community, is becoming physically weaker; nor is the prospect likely to be arrested."[36]

Frederick Hoffman's first publication, "Vital Statistics of the Negro" (1892) dealt directly with the issue of black extinction. Hoffman drew on secondary data gleaned from census—along with sociological, medical, and military sources—to make the argument that black health had precipitously declined following emancipation. Like many of his peers, he pointed to blacks declining share of the nation's total population—from 18.1% (in 1830) to 11.9% (in 1890).[37] Selective readings of the incomplete eleventh and twelfth censuses (1890, 1900) led many observers to predict that African Americans would become extinct inside of three to four generations. However, Hoffman conceded that the decrease in African American's share of the population, if not their actual decline in numbers, was due in large part to the massive influx of immigrants to the urban north, in contrast to the relatively immigrant poor South, home to 90% of the nation's black population. Subsequently, he confined his analysis to "native populations" (defined as those individuals who lived in their state of birth) of only five southern states: Alabama, Louisiana, Mississippi, and South Carolina. But in all five states the rate of natural increase—the excess of birth over deaths—occurred exclusively in the white as opposed to the black population. Hoffman concluded that, "something must be radically wrong in a constitution thus subject to decay. Even if he (blacks) be placed on equal grounds (to the white) he still will exhibit 'his race proclivity to disease and death'" seemingly confirming the Negro as impervious to even the most salutary environmental conditions.[38]

Actuaries defined mortality as the primary arbiter of individual and racial health. For Hoffman, mortality was nothing less than "the life story of a people."[39] Actuaries tracked mortality rates using "life tables," which indicated the probability of a person—for each year of their life—dying before their next birthday.[40] An analysis of life tables in Hoffman's early work, along with tables published in the journal of the *American Statistical Association* from 1892 to 1900, revealed a disproportionate rate of Negro and mulatto mortality with each successive year of their respective life spans. Further examination of white and black life tables in four northern, and four southern, cities from 1880 through 1890 revealed that the rate of black mortality was in greatest excess of whites under the age of fifteen—approximately 300%—and between the ages of fifteen and forty-five—approximately 220%. African American mortality also broke along gendered lines: under the age of twenty-five, mortality rates were much higher for males than females, a trend which continued without exception through all periods of life.[41] For Hoffman, black decline was rooted entirely in biology: "It is not in the conditions of life, but in the race traits and tendencies that we find the causes of excessive mortality."[42]

Hoffman's initial claims of black mortality—in both "Vital Stats" and later in *Race Traits*—were colored by the socioeconomic and cultural dynamics of Jim Crow. To gain an accurate sense of blacks natural increase—or lack thereof, birth rates had to be measured against corresponding death rates. Yet isolated rural blacks often failed to report births or deaths to the local white registrar, who in turn generally neglected to collect accurate data on resident black families. Statisticians tried to compensate by inferring the number of live births through a comparison of the consecutive decennial census. The distinction of "live births" is essential in that it did not account for the massive infant mortality rate among blacks.[43] Nevertheless, Hoffman asserted that it was "probably true" that the African American birth rate was more than whites, based on comparable data from the British West Indies and the northern states of Rhode Island, Connecticut, and Massachusetts, which, despite their small black populations, revealed a white birth rate nearly double that of blacks. African Americans' "gross immorality, early and excessive intercourse of the sexes" caused both unsanitary living conditions and high rates of venereal disease and tuberculosis among the race nationwide.[44] Hoffman's work on "Vital Statistics" shattered his belief in the transformative benefits of vocational education and led him to echo the most dire forecasts of black health: "The Negro, no matter how well educated, is not the mental or physical equal of the European, and the free "low-down" Negro will pass into a sure and deserved oblivion," thereby ensuing social progress.[45]

Across the social and natural sciences, the black body became a repository for a shifting constellation of various social ills—crime, sexual deviance, miscegenation and poverty—from which the wider white public had to be protected. Emergent ideologies of social biology—whereby social processes were both a cause and effect of biological imperatives—transformed blackness from an identity to a contagion that threatened the very fabric of American civilization. Beginning with "Vital Stats," Hoffman cast urban blacks as the prime purveyors of black degeneration. African Americans' idea that "freedom was free-er" in the towns and cities of the South stimulated a mass intra-regional migration which became particularly pronounced by the 1890s. From 1865 to 1895, the percentage of the black population of the sixteen largest southern cities had risen just over 242%, compared to an increase of 94% for whites.[46] This migration was seasonal in nature, characterized by the rhythms of the cotton harvests and the restraints of regional black codes. Hoffman defined black migration to the towns and cities of the New South as "one of the most distinct and deadly phenomena of the past thirty years." Even in centers such as Atlanta—showcase of the New South and home to a vibrant African American community, the blacks' death rate exceeded that of whites by 69%: 19.5 deaths per 1,000 for blacks, as opposed to 11.6 for whites. For Hoffman, "The loss of the farmer or planter will be the gain of the undertaker, for the drift of the Negro into the cities is usually a drift into an early grave."[47]

Actuarial Narratives of Crime and Race in Turn-of-the-Century America

At the turn of the century, amid rapid industrialization, urbanization and migration, blackness was primarily remade through crime statistics.[48] Accordingly, in Hoffman's second work, *Race Traits and Tendencies of the American Negro* published over two issues of the *Journal of American Economic Association* in 1896, crime and race took center stage. For Hoffman, the "intemperate habits" of black urbanites had turned large sections of the nation's cities into breeding grounds of disease, vice, and crime. Hoffman disregarded the myriad de jure and de facto segregationist policies such as anti-black real estate codes, municipal zoning ordinances, and curfews—which forced black migrants into the poorest and most decrepit areas of the nation's cities and towns. Instead, he sought to place the blame for urban blacks depravity in their genes. Hoffman argued that, in centers such as Philadelphia, Chicago, Boston, and Atlanta, the "large majority of the colored population is found to be living in the worst sections of the city, a section in which vice and crime are the only formative influences."[49] Hoffman concluded that "the colored race shows of all races, the most decided tendency towards crime in the large cities."[50] Despite sharing similar "conditions of life" with Italian, Russian, and Irish immigrants, black crime rates remained disproportionately in excess of their population. Hoffman's characterizations of black criminality as congenital or pathological in nature were informed by the work of the renowned late nineteenth-century Italian criminologist Cesare Lombroso. Yet whereas Lombroso cited sloping foreheads, narrowed eyes, or recessed chins as proof of criminality, Hoffman posited a black phenotype itself as a stigma of a diseased criminal nature. Blacks' depravity was rooted in the especially violent nature of their crimes: "As regards the most serious of all crimes, that of crimes against the person, the number of Negro criminals is out of all proportion to the numerical importance of the race."[51] Black prisoners were also disproportionately represented in national convictions for homicides (36.1%), and rape (40.88%). According to the census of 1890, there were approximately 82,239 prisoners in the United States, 24,000 of whom were black. Though blacks made up only 11.9% of the U.S. population, they comprised 29.18% of the national prison population. Nationwide, black men made up 38.21% of those charged with assault crimes. Hoffman believed that, if left to their own devices, blacks would become "more lazy, thriftless, and unreliable, until they will soon attain a condition of total depravity and utter worthlessness," as burdens on their local and the larger national community.[52]

Hoffman's analysis of black criminality revealed the contradictory logic of racial hereditarianism. Invariably, arguments for slavery's salutary effects ran counter to characterizations of former slaves and their progeny as innately and irrevocably degraded. Hoffman conceded that viable statistics for black crime during slavery were non-existent or fragmentary at best. Nevertheless, he maintained that following emancipation, blacks had chosen to reject "honest labor" for gambling, prostitution, and narcotics. For Hoffman, the "city negro's desire for finery" evinced his atavistic nature with often deadly consequences. The "roustabout" black male in his "pitiless search for employment honest or otherwise" often fell victim to "accidental death, frequent exposure to the inclemency of the weather, and not least his pronounced criminal tendencies." Young black men were far more likely to die a violent death, suffer injuries from gunshot wounds, knife fights, or "various other offrays" than their white counterparts nationwide: 19.5 deaths per 1,000 population for blacks, as opposed to 11.6 for whites.[53]

Sexual Deviance, Miscegenation, and Race Suicide in Turn-of-the-Century America

Hoffman's located his second metric of black extinction in the rising rates of venereal disease—specifically syphilis—and race mixing (miscegenation) within the nation's burgeoning urban black communities. Progressives held deeply contradictory views of African American sexuality, characterizing it as both violent threat and enfeebling trait. On the one hand, the rapid rise in lynching exposed southern anxieties about increasing black socioeconomic mobility and unregulated sexuality. During the 1890s, lynchings claimed some 139 lives each year, 75% of which were black. Litwack estimates that from 1890 to 1917, two to three black southerners were "burned, hanged or quietly murdered" each week.[54] Likewise, rising rates of venereal disease among urban blacks rarely failed to escape the attention of any observer of American race relations. Whereas in the country the vices of gambling, drinking and prostitution were tempered by social sanction and limited resources, the anonymity and opportunities of the city allowed such practices to flourish. From 1880 to 1890, black men living in cities such as New Orleans and Richmond were afflicted with syphilis at a rate almost triple that of their rural peers. In Atlanta, between 1890 and 1894, deaths from venereal diseases were 4.5 per 100,000 for whites, compared to 31 per 100,000 for blacks. During this same period, residents of Washington and Baltimore experienced almost identical racial disparities in venereal mortality rates.[55]

Race Traits cast the sexually promiscuous urban Negro as the author of his own destruction. Hoffman argued that "immense immorality was a race trait of which scrofula, consumption and syphilis are the inevitable consequences."[56] Syphilis in particular took a terrible toll on black bodies, resulting in skin lesions, respiratory problems, malformed limbs, and insanity. The disheveled and mentally ill syphilitic—an increasingly common sight in the black sections of many southern towns and cities—seemingly embodied the deleterious effects of city life on blacks. Nor were whites alone in condemning the sexual deviance of urban blacks. A black minister from Montgomery, Alabama echoed such anxieties about the depravity engendered by urban life: "In the lynchings I have known about, the victims were always men in the community no one could say a good word for. They came out of the slums at night, like the diseased raccoon and stole back again."[57] For observers across the color line, blacks' seeming inability to tame their sexual urges destroyed their bodies, threatened respectable society, and revealed their inability to internalize the work ethic of modern industrial life.

Black female sexuality also came under intense scrutiny by Hoffman. Historians have noted how popular understandings of evolutionary theory and sex selection assigned women a clearly defined role in racial uplift, which could "make or break a race."[58] Hoffman claimed that the high rates of venereal diseases among blacks were due to an "immense amount of immorality, which is a race trait and of which scrofula and syphilis are the consequences." He cited much of this immorality in the "enormous waste of child life engendered by the unchaste women of the race whose bastard progeny" continued the cycle of sexual and social depravity.[59] Though Janus faced characterizations of black female sexuality—the asexual Mammy and ravenous Jezebel—had long infused American society, Progressive Era definitions of this dichotomy were deeply eugenic in nature. Reformer Eleanor Taylor, writing in *Outlook Magazine*, cited black women's "moral decadence" as detrimental to racial health, arguing that "it [is] her hand that rocks the cradle in which the little pickaninny sleeps."[60] Throughout *Race Traits*, Hoffman sought to give statistical coherence—via birth, infant mortality, and venereal disease rates—to the perceived social threats posed by the unrestrained sexual practices of the emerging black proletariat.

Race Traits painted a picture of a rapidly urbanizing Negro beset by the twin demons of venereal disease and race mixing. Hoffman believed racial health was synonymous with racial purity. Only the inferior degenerate races failed to protect their racial integrity: "It may be said, only with emphasis, that the crossbreed of white men and colored women is, as a rule, a product inferior to both parents, physically and morally."[61] Hoffman, like many of his peers across the social and natural sciences, subscribed to the "law of similarity"—the idea that like produced like, and that "affection between groups and individuals was based in large part on sympathy."[62] These various forms of sympathy developed out of similar interests, ideas, and habits that bound social groups together. Conversely, a lack of sympathy seemingly made affections, and thus reproduction, socially and biologically impossible. Yet some thinkers believed that race mixing would help blacks ward off extinction through a much-needed infusion of white blood.[63] In 1892, Joseph LeConte, the South's most prominent natural scientist, proposed an intriguing alternative to the seemingly inevitable fate of the Negro's extinction: "At present for the lower races everywhere there is eventually but one of two alternatives: either extermination or mixture." LeConte believed that while the mixing of extreme types such as the "Teuton and the Negro would produce only the worst results," amalgamation between "marginal varieties of the primary races" could perhaps serve to stave off extinction for the lower races. Nevertheless, observers across the color line generally concurred that race mixing had a deleterious effect on both individuals and the republican body politic.[64]

Despite the presumed social and biological imperatives of the "law of similarity," individuals inevitably strayed across the color line with great regularity. For Hoffman, black urbanization had led to a direct increase in the general population of "mulattos," or peoples of mixed race. Census data showed an increase of mulattos in the black population, from 12.0% (584,000) in 1870 to 15.2% (1,132,000) in 1890.[65] Confronted with the seemingly contradictory evidence of increased race mixing, Hoffman maintained that hereditarian imperatives punished the practice by reducing the ability of hybrid offspring to reproduce. He argued that mixed race unions produced an average of only one child, compared to 2.8 children born to intra-racial couples.[66] Hoffman cited the editors of *Spectator* who claimed that "the mortality of the Negro may well be considered the most important phase of the so-called race problem; for it is a fact which can and will be demonstrated by indisputable evidence, that of all races for which statistics are obtainable . . . the negro and in particular his hybrid character, shows the least power of resistance in the struggle for life."[67] Through the metric of mortality, prevailing notions of infertile mulattos were reconfigured along monetary lines: race mixing led to a drop in the race's natural increase, which in turn devalued black lives as insurable commodities.

The figure of the mulatto was a harbinger of racial decline in the progressive imagination. With the rise of hereditarian thought, popular understandings of "tragic mulattos" torn between two cultures slowly gave way to notions of race mixing as both a social and biological perversion. Theories of black inferiority were predicated on notions of the black body and mind as instinctively primal. Professor J. F. Miller of Columbia University credited an increase in the number of blacks in southern lunatic asylums to "the modern influences and agencies on his less developed nervous organization."[68] Mental illness in African Americans was seen as the result of a primitive physiology taxed beyond both its physical and mental limits by the demands of modern life. The degeneracy of the mulatto, "imprisoned by two warring constitutions," stemmed from a near fatal imbalance in which a civilized white mind was enfeebled by a savage black body. Blacks' high rates of venereal diseases, coupled with a small but significant infecund portion of their hybrid population, meant that the "negro race was lagging, and sure to expire" in the zero-sum race of evolution. Late nineteenth-century debates over "race mixing" and interracial marriage also divided the black community between those who came to view it as the surest path to social, economic, and biological equality—such as Frederick Douglass—and those who believed that despite the need for its decriminalization, rampant intermixture would lead to racial extinction—such as W. Calvin Chase, the editor of the influential *Washington Bee*, who argued that race pride and racial health were contingent on the maintenance of social, cultural, and biological racial integrity.[69]

"The Negro Lung": Vital Capacity as an Index of Racial Fitness

Hoffman's third and final metric of black extinction was based in the race's apparent "diminished vital capacity." Hoffman argued that blacks' innate criminality, sexual depravity, and high rates of racial admixture "made them unequal to whites in their power to resist disease."[70] Turn-of-the-century actuaries used the category of "vital capacity" to denote an individual's respiratory health: lung capacity, circulation, and resistance to pulmonary diseases such as tuberculosis. But "vital capacity" also had far reaching socioeconomic and cultural connotations via which social scientists "constructed a model of work and the working body as pure performance, as an economy of energy," according to historian Anson Rabinbach.[71] The metaphor of the human motor—informed by the emerging theory of thermodynamics—framed the working body as a self-contained dynamo locked in a constant struggle to transcend the human element expressed through the indices of "fatigue."[72] Physicians and social scientists measured the health and monetary value of working bodies by determining the rate at which they could efficiently labor before succumbing to fatigue. Racial fitness was predicated on stamina and the endurance to labor: the fit races were those whose physiologies were less susceptible to debilitating diseases such as tuberculosis, whereas the unfit were defined by their high rates of respiratory diseases and congenital incapacity for the rigors of modern industrial work.

Respiratory health was based on a number of factors: lung capacity, chest circumference, and height-to-weight ratio. Spirometry—the primary medium for measuring respiratory health in the nineteenth century—was employed for the first time on a major scale by the Union army during the Civil War.[73] In his research for *Race Traits*, Hoffman drew heavily on the wartime studies of Col. Benjamin Gould for analysis of disparities in racial respiratory health. Though Gould found minimal differences in chest circumference—35.1 inches for black and 35.8 inches for white recruits, the disparity in lung capacity was especially striking. Army medics measured the lung capacity for whites at 184.7 cubic inches, 163.5 for blacks, and only 158.9 for mulattos.[74] Gould also found that whites possessed higher rates of respiration than their black or mulatto counterparts. Hoffman argued that during the antebellum era, blacks' "innate lack of exertion" led them to squander opportunities in the mechanical arts afforded to them by their kindly masters. And while Hoffman found that blacks did possess "latent mechanical aptitude," an inferior vital capacity and lack of "civilized stamina" ultimately hindered their industrial progress.[75]

Hoffman remained convinced that the source of blacks' inferior respiratory health as well as their disproportionately high rates of tuberculosis lay in their low height-to-weight ratio. Although low body weight was not seen as a direct causal agent of respiratory diseases such as tuberculosis, it was an indicator of possible infection. Yet Hoffman shied away from drawing a direct correlation between low body weight and tuberculosis, stating, "The uniform result of statistical investigations of life insurance companies has proven that persons under average weight have a decided tendency toward pulmonary diseases."[76] An 1895 study by Prudential seemingly confirmed that a low body weight-to-height ratio was a determining factor in consumption. To Hoffman's surprise, blacks possessed a higher weight-to-height ratio than whites. Nevertheless, he remained adamant—even in the face of blacks' superior rate of respiration, approximate chest measurement, and average body weight— that the "lung capacity of the colored race is in itself proof of an inferior physical organism" on both an individual and group level.[77]

Tuberculosis had a complex etiology rife with race and class connotations. Robert Koch's 1883 isolation of the tuberculosis bacillus reshaped the ancient affliction of consumption in the etiological terms of germ theory. Subsequently, tuberculosis was understood as a product of environmental factors that created and or exacerbated an individual's predisposition to the disease. Koch's discovery occurred during a time of rapid industrialization in which cities became inundated with masses of the working poor. In sociocultural terms, tuberculosis was characterized as a contagion resulting from the congestion, poverty, and filth that accompanied the rapid rise of modern civilization. Katherine Ott notes how, by the late nineteenth century, consumption was redefined as a disease of "over-civilization" afflicting the enervated and neurasthenic middle and upper classes. While the ravages of consumption spared no section of society, it was the working classes who suffered most from the dreaded white plague.[78] Yet, informed by the imperatives and mores of contemporary political economy, the analysis of the tubercular poor cited their genes—rather than their environment—as the primary culprit of their condition.

The high rates of tuberculosis among blacks did not fail to escape the notice of the era's army of self-professed "race experts." Hoffman drew on a smattering of urban health records and decades-old Civil War medical evaluations of Union recruits to advance his thesis of a race ravaged by tuberculosis. Civil War records revealed that while whites had been rejected for consumption at a rate of 11.4 per 1,000, black rejection rates were substantially lower at 4.2 per 1,000. Hoffman maintained that while mortality rates from consumption had been relatively equal between the races during the antebellum era, the years since emancipation had witnessed a dramatic rise in black mortality rates. He buttressed this theory with data from the 1890 Census and an examination of tuberculosis rates per 100,000 of the fourteen cities nationwide with the largest black populations. Without exception, urban mortality rates for blacks far exceeded those of whites. Even in northern centers such as Boston and New York, which had relatively small black populations, the racial disparities in tubercular mortality rates were staggering: in New York, black mortality rates were almost three times the rates of whites.[79]

Blacks' apparent diminished respiratory health was seen as both a cause and effect of their downward evolutionary trajectory. Though Hoffman was willing to see slavery as an ameliorative social force, he was unable or unwilling to conceive of the negative environmental effects of urban life on the black working poor. For Hoffman, blacks' innately degraded habits—not their environment—spawned the squalor that allowed the tuberculosis bacillus to thrive. Blacks' seemingly undisciplined and diseased nature therefore necessitated their expulsion from the nation's labor economy. The arrival of rural blacks in southern cities placed added strain on a job market already weathering a deep nationwide depression. Within this socioeconomic context, tuberculosis functioned as both an etiological condition and an imperative of a white supremacist political economy. The ever prescient DuBois noted this shift in tuberculosis, from an immigrant disease to a black disease at the century's end: "Negroes are not the first people who have been claimed as its particular victims; the Irish were once thought to be doomed by that disease—but that was when Irishmen were unpopular."[80] The increasingly malleable boundaries of whiteness allowed groups such as the Irish—and to a lesser extent the Jews—to shed their identities as consumptive races and gradually attain the healthful character of republican manhood.[81] In contrast, observers such as Dr. F. Billings of the U.S. Army maintained that "the Negro's extreme liability to consumption alone would suffice to seal his fate as a race." As one southern public health official bluntly stated, "Disease is today almost a synonym for the word 'Negro.'"[82] For the era's managerial elites, the drive to deny blacks full citizenship and relegate them to the margins of the labor market was driven by both social and biological imperatives.

Depictions of African Americans as a consumptive race lacking in vital capacity were driven by the socioeconomic, cultural, and spatial dynamics of the Progressive Era political economy. In 1880, approximately 80% of blacks lived in the states of the former confederacy. Black labor and capital, even in their marginalized forms, was key to the economy of the New South. Indebted black sharecroppers were crucial fixtures in the rural economy and unskilled black labor formed the backbone of the regional iron and textile industries.[83] Hoffman's actuarial assessments of black bodies as defective commodities reinforced their subordinate position within the southern labor economy. These actuarial narratives of black inferiority also helped rationalize organized labor's nationwide hostility to the black worker as "a cheap liver who demands less wages" due to his "natural proclivity for filth and debasement."[84] Hoffman believed that evolution had engineered specific bodies for specific environments. With the ascendance of Mendelian biology—which privileged racial traits over acquired characteristics, removing a race from its natural habitat led to a marked decrease in vital capacity and possible death. He insisted that blacks had experienced irrevocable long-term physiological damage by way of their removal from Africa and subsequent manumission from a benevolent and nurturing slavery. From both a spatial and evolutionary perspective, working black bodies were out of space and out of time.

Racially informed climatology reimagined the respiratory tract into a key site of racial labor division. The rubric of vital capacity became a key technology of both domestic and imperial political economy. Hoffman attributed blacks' lack of respiratory health and their small lung size to their "tropical heritage" in sub-Saharan Africa. While the "arctic lung" of northern European whites required a large oxygen capacity to maintain animal heat in cold regions, "it would be in accordance with the economy of nature to suppose that the oxygen capacity of a tropical lung would be smaller than that of the arctic in the same ratio to maintain animal heat in the sultry climate of the Equator."[85] Hoffman cited the work of Ira Russell, a member of the Civil War-era U.S. Sanitary Commission who had studied racial lung weights. Based on an indeterminate sample Russell had found, the average weight of the "negro lung" was approximately four ounces less than the lung of whites. Russell's work on racial respiratory health remained the definitive model for decades across both the natural and social sciences.[86]

America's turn-of-the-century imperial interventions throughout the Pacific and the Caribbean introduced its social and natural scientists to a transnational discourse of "climatology" that posited climate as the determining factor in racial evolution. Much like their European predecessors, American colonial officials worried over the ability of white bodies to adapt and thrive in tropical climates. Sir George William Des Voeux, a British colonial administrator echoed the concerns of many Americans when he expressed his "grave doubts [about] whether any tropical country can became a prosperous white man's colony . . . where white men are laborers as well as employers, able to rear a healthy progeny, inclined to and physically able of work with the hands."[87] American officials such as Col. George Gorgas and Col. Leonard Wood conducted extensive studies of American troops in Southeast Asia and the Caribbean, measuring their susceptibility to diseases such as malaria and yellow fever.[88]

Medical and military personal were also deeply troubled by a pervasive yet indefinable "lassitude" that seemingly took hold of white men who spent extensive time in the tropics. Warwick Anderson notes that fears over the tropics as a "white man's graveyard" led to a revaluation of the interconnections between manhood, race, and labor in the nation's rapidly expanding political economy.[89] Whether black men leaving the rural south or white men pursuing the "white man's burden" abroad, both seemed to suffer the enfeebling physiological effects—such as a decline in respiratory health—of leaving an industrial civilization run amok. Models of the tropical and arctic lung sought to reconcile racial form with function and affected racial labor division along distinct spatial lines.[90]

Ultimately, actuarial narratives of racial fitness were used by social scientists to rationalize production in both the public and private spheres at home and abroad. Martha Banta describes this utopian impulse as the broader societal pursuit of a "managed life." Banta argues that modern cultures of management efficiency, on and off the shop floor, reified existing gender and racial hierarchies while producing new models of articulating these respective forms of social difference. The logic of turn-of-the-century management theory maintained that subaltern bodies such as women and non-whites represented the "wasteful intractability of the savage element . . . introducing 'wild facts' into a situation where managers feared unmanaged, unpredictable irrationality above all else."[91] The racial demands of Progressive Era management theory ensured that the rhetoric of social efficiency was a language of whiteness. The category of vital capacity constructed by industrialists, scientists, physicians, social scientists, and actuaries allowed for the quantification of bodies which otherwise defied quantification.

Race Traits achieved immediate notoriety following its publication in spring 1896. Hoffman's work spread in popularity far beyond the narrow confines of actuarial science and became one of the era's definitive texts on race. White intellectuals greeted it with near unanimous praise for its seemingly objective analysis of the "ever vexing Negro problem."[92] Professor W. F. Blackman of Yale University claimed that, "In dealing with the Negro question we have had enough of assumption, prejudice, sentiment and timidity; what we need is exact research in accordance with the methods of anthropology and statistics provided by Mr. Hoffman."[93] The biologist F. Lamson Scribner commended Hoffman for the way in which his conclusions were "intelligently and impartially combined and secured in a clear and attractive manner." Reviewers argued that unlike previous "amateurish" and "rank scientific" racial analysis, Hoffman had produced a work of ineluctable scientific fact.[94] Professor W. B. Smith of Tulane University chillingly noted Hoffman's work as a reminder that evolutionary and national progress depended on the death of the Negro: "The vision of a race vanishing before its superior is not at all dispiriting, but inspiring."[95]

The lavish praise accorded to *Race Traits* was invariably linked in part to the foreignness of its author. Indeed, the editors of *Dial* magazine hailed it as, "a thoughtful work by an unbiased foreigner" and an essential read for all serious students of the race problem.[96] As an immigrant, Hoffman posed as the consummate outsider, a dispassionate observer of race relations, unburdened by native-born Americans racial bias and prejudices. Moreover, as a German immigrant, Hoffman constantly played upon the American intelligentsia's reverence for Germanic philosophy and models of higher education. Bismarckian Germany was at the center of progressive transatlantic networks of social theory and a beacon of reform for educated Americans across the color line. Hoffman was merely one of the first in a long line—culminating almost a half-century later in Gunnar Myrdal's *American Dilemma*—of seemingly objective white European observers of modern American race relations whose foreignness and ability to dispassionately reduce the Negro to an object of social scientific inquiry, confirmed on them the status of experts.[97]

The reception of *Race Traits* among the black intelligentsia was decidedly mixed. Reactions to *Race Traits* came from two primary sources: black intellectuals and the fledgling black insurance industry. W. E. B. DuBois, an ambivalent exponent of evolutionary theory, praised Hoffman for his exhaustive research while nevertheless taking him to task for "shoddy methodology" and improper "application of the statistical method." DuBois argued that Hoffman had failed "to see that such a method is after all nothing but the application of logic to counting, and that no amount of counting will justify a departure from the severe rules of correct reasoning."[98] DuBois, a devout Germanophile, contended that Hoffman's refined continental methodology and rules of reasoning had been compromised by his co-option of American racial biases and preconceptions.[99] For DuBois, rational inquiry and scientific empiricism were immune to the coarse subjectivities of racism. Yet even such an adept social critic such as DuBois did not dispute Hoffman's privilege or intent in counting, only the manner in which he had done so. A few years later, DuBois would adopt similar statistical methodologies in his seminal sociological analysis, The Philadelphia Negro (1899). However, he would take great care to distinguish his analysis of black criminality and health from popular interpretations such as Hoffman's, which linked these tendencies to biological characteristics.[100]

African American critiques of actuarial narratives of race suicide were hampered by the predominance of empirical inquiry in Progressive Era discourse. The dominance of the scientific method in both the social and natural sciences—a focus on processes rather than content—divested black intellectuals of the epistemological tools needed to challenge Hoffman's theory of race suicide. DuBois and his peers were forced to focus on how Hoffman had constructed his evidence, rather than the ideological and conceptual frameworks that had informed his research. Kelly Miller, a leading black sociologist and mathematician, was initially effusive in his praise of Race Traits, declaring it the most "important utterance on the subject of race since the publication of Uncle Tom's Cabin."[101] However, Miller soon began to vigorously dispute Hoffman's assertion that black health had flourished under slavery, only to decline following emancipation. Miller noted that in the previous eight decades, the black population had quintupled with little aid from external migration: "How a people who had shown such physical vitality for so long a period, has all at once, in the past decade become relatively infecund and threatened with extinction?" The answer did not lie in any innate racial traits or tendencies but in blacks' "various conditions of life."[102] Convinced that "a sound and objective application" of statistical analysis of black life would yield an accurate picture of the race's fitness, Miller offered the following advice to blacks: "To the Negro I would say, let him not be discouraged at the ugly facts which confront him . . . the Negro should accept the facts with becoming humility and strive to live in closer conformity with the requirements of human and divine law. He does not labor under a destiny of death from which there is no escape."[103] Couched in the rhetoric of racial uplift and respectability, Miller fashioned a cogent critique of biological theories of black degeneracy within the bounds of rational inquiry.

The African American business community responded to actuarial narratives of the race's decline with entrepreneurial zeal. Shortly after the publication of Race Traits in spring 1896—and the subsequent segregation of the insurance industry, a small group of black businessmen moved to reorganize local fraternal aid societies along a corporate model. The first black insurance company formed on this basis was the National Benefit

Insurance Company, founded in 1898 by S. W. Rutherford in Washington, D.C. That same year, C. C. Spaulding and a consortium of black businessmen organized the North Carolina Mutual Benefit Insurance Company. One black businessman thanked Hoffman for discouraging white firms from insuring blacks: "Without Mr. Hoffman's pernicious propaganda, there would have been no North Carolina Mutual."[104] In 1908, Herman Perry of Atlanta, Georgia began organizing Standard Life, which would soon become the largest black life insurance company in the world and a beacon of racial uplift. Even so, Standard Life was not fully incorporated until 1913 after Perry waged a protracted battle to raise the sufficient funds set by the state authorities to start up the business.[105]

Notwithstanding the petit bourgeois ethic of black insurance companies, their very existence radically challenged the constraints of a segregated political economy. Black-owned insurance companies were some of the first institutions in the New South—and the nation as a whole—to allow the race to acquire both financial and social capital. DuBois cited the subversive character of black entrepreneurism, noting that while white southerners "feared Negro crime, they feared Negro success and ambition more . . . the south can conceive neither machinery nor place for the educated, self-reliant, self-assertive black man."[106] Such assertiveness was evinced in industry literature, which advised black agents: "We cannot write any white insurance business and the white agent is controlling the insurance situation in our homes. This will never do! Break it up! Get the Business! Be enthusiastic, alive to the task; hard driving, never satisfied until you get the business."[107] Kevin Gaines describes the era's black middle class as a "group bound by a shared ideological preoccupation with bourgeois status, rather than one sharing in the material benefits commensurate with the white middle class." [108] Black uplift ideology formed the basis for a racial elite identity, which equated class stratification with racial progress.

Frederick Hoffman's early work in "Vital Statistics of the Negro" and *Race Traits* was essential in maintaining African Americans' exclusion from the mainstream insurance industry—from the late nineteenth century until the eve of the Second World War. In 1940, the vast majority of white underwriters still refused to insure blacks and of the fifty-five firms that did, only five did so at standard rates. Hoffman was quite vocal about his role in persuading the nation's major insurers to exclude blacks, "in light of my prior work in charting blacks' excessive mortality and debased character traits, Prudential has long since stopped soliciting risk policies from the colored population."[109] However, in the face of the emergent civil rights movement—born in part by the struggle against fascism abroad—white resistance to insuring blacks began to fade in the postwar era. By 1957, over one hundred white companies competed for black policyholders at standard rates. By the early 1960s, Congress admitted black insurance companies such as North Carolina Mutual into the select pool of companies underwriting group life coverage for federal employees and military personal. Finally, in 1962, the American Life Convention and the Life Insurance Association of America, the lily-white paragons of the life insurance trade associations, made the Mutual their first black member.[110]

Race Traits spawned a veritable cottage industry in black race suicide literature. The early twentieth century saw a surge in works predicting blacks' imminent demise—from Charles Carroll's *The Negro a Beast; or in the Image of God?* (1900), W. P. Calhoun's *The Caucasian and the Negro in the United States* (1902), W. B. Smith's *The Color Line: A Brief in Behalf of the Unborn* (1905), and Robert Shufeldt's subtle *The Negro, A Menace to American Civilization* (1907).[111] In 1907, historian Edward Eggleston published his chilling screed, *The Ultimate Solution of the American Negro Problem*, in which he contrasted the robust body of "the darkey under slavery" with the "diseased carriage of today's negro vagrant" as proof of blacks resurgent atavistic character.[112] Eggleston cited Hoffman's "unerring statistical proof" as evidence that the "natural flow of evolution would soon dispense with the negro."[113] Dr. Paul Barringer, Chairman of the Faculty at the University of Virginia, and author of *The American Negro: His Past and Future* (1900), referenced Hoffman at length in arguing for "the Negro's generic tendency to enfeebling savagery."[114] Barringer believed that, for whites, "the Negro Problem rises above a question of altruism and becomes a question of self-preservation."[115] Hoffman hoped his work would serve as a clear "condemnation of modern philanthropic attempts of superior races to lift inferior races to their own elevated position," a process so futile that "it would seem criminal indifference on the part of civilized people to ignore it."[116] Spectral images of degenerate Negroes also took on literary forms in contemporary popular culture. Thomas Dixon Jr.'s Reconstruction trilogy—*The Leopard's Spots* (1902), *The Clansmen* (1905) and *The Traitor* (1907)—was key in popularizing notions of blacks as "crazed and monstrous beasts" utterly unsuited to the demands of work and citizenship in a modern industrial republic.[117]

Following the publication of *Race Traits*, Hoffman lobbied his superiors at Prudential to fund a transnational investigation of racial health and mortality. Through the metrics of criminality, venereal disease, miscegenation, and vital capacity, he sought to connect the domestic "Negro problem" to the global race problem—and gain a clearer understanding of the "vicissitudes of racial health on a global scale," which had previously been "tainted by race prejudice causing even the most fair minded observer to err in judgment."[118] Hoffman's studies of "the constitution of the color line" took him through the American southwest, Caribbean, Latin America, and South Pacific—all territories acquired by the U.S. in the previous quarter-century and subsequently integrated into America's rapidly growing imperial political economy.

In 1915, Hoffman arrived in Hawaii—which had been annexed by the U.S. in 1898—to survey the feasibility of insuring the islands' polyglot population. As late as 1901, the editors of the *Honolulu Commercial Advertiser*, citing excessive criminality, race mixing, opium abuse, and the spread of "loathsome diseases" such as tuberculosis, had predicted the demise of the island native inhabitants. The editors claimed, "There are probably now living men of voting age who will see the last full blooded Hawaiian native—there is something tragic in the utter annihilation of a race, especially one so amiable in many respects as are the Hawaiians and it is to be hoped that something may be done to check the tendencies that are causing the decay."[119] Upon arrival in the islands, Hoffman was immediately struck and disturbed by the prevalence of race mixing among the natives, which had produced "an envenerated race beset by poverty, drug use, and an 'impotent lassitude.'" All of the vices of the uncivilized were present: "The neglect of infants by native mothers, the gradual diminution of the fish supply, and the spread of loathsome disease [tuberculosis] have been reinforced by the newly-acquired vice of opium smoking." Moreover, the mixed Hawaiians "lacked the racial vitality and self-control of the Asiatic consumers and the results of this vice are therefore the more disastrous."[120] Hoffman's travels in the Pacific and the West Indies seemed to support the prevailing belief of Progressive Era social scientists in the primacy of race and racial admixture "in determining an individual's life span and resistance to disease."[121] Hoffman's sense of racial noblesse oblige led him to conclude that although "the colored man was not nor would ever be the equal of the white man," a "just sense of international race relations depended upon the precise recognition of inherent and permanent inequality in physical, mental and moral traits"—the delineation of racial difference being seen as key to long-term racial equality whether on the islands or stateside.[122]

Following extensive actuarial surveys, Hoffman informed Prudential's president John S. Dryden that contrary to prevailing wisdom, the Hawaiian race was not doomed to excitation. Hoffman acknowledged native Hawaiians' "predilection for degeneracy, before citing how a benign climate, racial intermixture and (healthful) plantation economy made Hawaii "perhaps the most conspicuous modern illustration of successful tropical adaption and race progress."[123] He juxtaposed Hawaii's "racial progress" with racial conditions stateside: "The Negro death rate remains considerably in excess of the white death rate, regardless of far reaching sanitary improvements of more or less similar effect upon both populations."[124] Hoffman's efforts to reconcile these disparate actuarial narratives of racial health were firmly situated in the transition from bonded to contract labor on the part of the nation's oldest and newest sets of colored laborers. Whereas black degeneration was the result of their migration out of the plantation economies of slavery and sharecropping, Hawaiians' salvation lay in their ever-increasing immersion into the fruit and sugar plantations of corporations such as Dole and United Fruit, which, in both cases, the plantation became a key site of race management.[125]

The racial dynamics of health—and race relations in general—remained a fascination of Hoffman throughout his long career as an actuary and public intellectual. Though Hoffman never wavered in his belief of Negro inferiority, by the mid-1920s he had revised his earlier theories regarding blacks' extinction, arguing that blacks would more likely assume a "stagnant position much like the American Indian and the Gypsies of Europe."[126] Hoffman retained a strong sense of racial paternalism—forever regarding himself as a "friend to the Negro"—claiming that although "the colored man was not nor would ever be the equal of the white man, a just sense of international race relations depended upon the precise recognition of inherent and permanent inequality in physical, mental and moral traits."[127] Like many of his peers, Hoffman saw no contradiction in his vision of an American pluralism built on white supremacist racial hierarchies. His fidelity to empiricism and the seemingly ineluctable facts of statistical science blinded him to the racialist imperatives in his research methodologies. Nancy Stepans reminds us that "the scientists who gave scientific racism its credibility and respectability were often first rate scientists struggling to understand what appeared to them to be deeply puzzling problems of biology and human society" and would have dismissed any form of overt prejudice or "race hatred" as quite beside the point.[128] A set of circumstances that clearly reveals the extent to which racialist thought was inextricably linked to the era's empirical methodologies: the delineation of difference was merely an objective process—therefore immune to coarse racial prejudices.

Extinction discourse remained a powerful explanatory model of African American proletarianization well into the interwar years. Actuarial narratives of African American race suicide provide key insights into the persistent influence of non-economic factors such as race on Progressive Era political economy. Statistics and actuarial science allowed progressives to chart the precise rate and value of blacks' transition to industrial modernity. White elites believed that mastery of this racial knowledge was crucial to the development of a rational and efficient social order. Susan Schweik, in her analysis of early twentieth-century anti-vagrancy laws that criminalized the physically disabled in public spaces, details how race has long played a role in the "symbolic economy of disease."[129] Throughout the nation, "ugly laws," while primarily aimed at the physically disabled, were also used as a key form of racial labor division. Schweik notes that the "concept of disease has long been tied to racial hierarchies, and the barrage of statistics brought forth in the name of socio-medical racism in the late nineteenth and early twentieth centuries hammered home the point that blacks posed a major health and social menace."[130] Black leaders were not immune to this mindset, and constantly evoked the race's social and biological value. As a guest at the First National Conference on Race Betterment, Booker T. Washington inveighed against this rising eugenic tide: "The American Negro is worth saving, and making a strong, helpful part of the American body politic."[131] Washington was well aware that any calls for inclusion in the body politic would have to conform to the racialist logic of the day and foreground the corporeal dimensions.

The demands of rapid industrialization, imperialism, and mass consumerism required the creation of a "vanishing Negro." Actuarial science provided the conceptual and practical means to identify, evaluate, and monetize key demographic shifts in the nation's political economy. Historians have described statistics as a "technology of distance" well suited to conveying concise, accurate, and easily understood information about ambiguous subjects such as racial identities.[132] Hoffman's narratives of black health were ideological constructions—exercises in race making—that allowed those who had never met an African American to gain a working knowledge of one. Megan Wolff notes that "the claims Hoffman made about his own methodology and the access it provided to complete knowledge—the promise that he could examine a series of numbers and know, infallibly, the truth—invoked the 'myth of the actuary' as well as its underlying myth of progress."[133] As a professional actuary, writes Wolff, "Hoffman stood at the intersection of two related discourses: progress and profit," which represented a meeting of the social Darwinism of the marketplace with the social Darwinism of racial ideology.[134] Actuarial narratives of race suicide linked ostensibly premodern traditions and peoples, such as blacks, to qualitative and quantitative forms of modern labor economy, such as statistics, to reveal the societal and physical perils of race, work, and the very limits of progress amid the shifting landscape of modern industrial America. Yet with the coming of war in Europe and its reverberations stateside, the figure of the Negro, which heretofore had served as a regional—generally southern—metaphor for the racial perils of progress, now became in both theory and practice a national industrial agent: the "Negro problem" was now an American problem.

2

The Negro Is Plastic

The Department of Negro Economics, Sociology, and the Wartime Black Worker

"The wonderful metamorphosis now in process, by which cotton-pickers are being transformed into steel workers, is quite as interesting and has as many approaches and slants toward industrial economy as any event that has occurred in our history."
—Philip H. Brown, *The Negro Migrant* (1923)

"There is no such thing as economic growth which is not at the same time, growth or change of a culture; and the growth of social consciousness, like the growth of a poet's mind, can never, in the last analysis, be planned."
—E. P. Thompson, *Customs in Common: Studies in Traditional Popular Culture*

In July 1917, less than four months after the United States entered the Great War to make the world safe for democracy, the whites of East St. Louis, Illinois massacred dozens of their fellow Americans. White workers—incensed by increased job competition and the use of black strike breakers at the local packing plant the previous summer—exploded with rage when rumors spread of black men socializing with white women at a recent union gathering. For three sweltering days in early July, a mob of whites clubbed, beat, shot, and burned alive close to 125 black men, women, and children. As one Congressman remarked at the time, "It is impossible for any human being to describe the ferocity and brutality of that mob." Surveying the aftermath, Oscar Leonard, a superintendent at the local Jewish Educational and Charitable Association, observed: "When I went to East St. Louis to view the sections where the 'riots' had taken place, I was struck that the makers of Russian pogroms could learn a great deal from the American rioters. . . . The Russians at least gave the Jews a chance to run while they were trying to murder them. The whites in East St. Louis fired the homes of black folk and either did not allow them to leave the burning houses or shot them the moment they dared to escape the flames." The ferocity of the rioters shocked even the most hardened observer of American race relations.[1] On July 28, 1917, ten thousand black New Yorkers marched through Harlem in silent protest against the massacre in East St. Louis. Marchers brandished banners that read, "We are Maligned as Lazy and Murdered when We Work," "Patriotism and Loyalty Presuppose Protection and Liberty," and "Give Us a Chance to Love our Country."[2] Protestors linked themes of race, labor, and citizenship to the rhetoric of wartime patriotism—to push the "race problem" into the wider national consciousness for the first time since Reconstruction.

Progressives' steadfast belief in empirical inquiry as a means to social reform meant that no "problem" was without a solution. Indeed, in the wake of Reconstruction, most northern whites had been rather ambivalent about racial violence, dismissing it as a macabre southern fetish. Most northerners saw the tensions between capital and labor—the so-called labor problem—as one of the biggest challenges facing the nation. With the coming of the war, however, the increased competition between recent European immigrants, native-born whites, and southern black migrants over jobs, housing, transportation, and education made race relations an increasingly common "problem" in northern and Midwestern cities such as East St. Louis.[3] Race and race relations subsequently became popular fields of study in both the natural and social sciences in research institutions across the nation. Moreover, the war gave theorists of industrial evolution—those who viewed industry as an expression of racial fitness—the opportunity to reconfigure social policy along biological lines. The fledgling discipline of sociology—with its focus on social behaviors and institutions—was well suited to make sense of blacks' transition into industrial modernity. In the weeks following the violence in Illinois, Assistant Secretary of Labor Louis Post called on Washington to redouble its efforts in investigating the racial dynamics of wartime labor to "forestall further violence as recently witnessed in East St Louis."[4] Wartime imperatives required rational expertise endowed with federal authority to ease the spasms of bloodlust from a populace struggling to come to terms with the rapidly shifting demographics of a wartime labor economy. Sociologists were instrumental in framing the terms of this debate and, perhaps more than any other discipline, turned the Negro into a national figure of social scientific inquiry.[5]

World War I occasioned a shift from hereditarianism to cultural models of the "Negro problem." The Negro of the prewar era, despite being overwhelmingly confined to the South, had been a "known" figure in the wider progressive imagination. The Negro was both an active threat to American civilization or a pitiable figure doomed for extinction in the brave new world of American industrial modernity. Northern observers gleaned their knowledge of blacks and black life from self-appointed southern race experts and popular culture. A burgeoning mass culture centered in the North transmitted largely southern characterizations of child-like, buffoonish, or rapacious "darkies," through song, stage, and the page to an appreciative nation.[6] But following the mass wartime exodus of blacks to the North, these mediums became largely ineffectual for making sense of rapidly shifting racial labor dynamics. Despite the increased physical presence of blacks in the North, the Negro remained something of a mystery—which an eclectic group of thinkers across the natural, social, and artistic realms sought to solve.

The wartime state was both a catalyst and agent of black proletarianization. Using the legislative and judicial mechanisms of the state progressives hoped to restructure the nation's labor economy along racial lines. Whereas the exigencies of war—cessation of European immigration and industrial labor shortages—drove blacks north, the state also actively sought to transform blacks into a modern workforce. Analysis of the Department of Negro Economics (DNE)—the first federal agency devoted exclusively to black labor since Reconstruction—reveals a key attempt on the part of reformers to reconcile the "Negro problem" with the "Labor problem" within the institutional nexus of early twentieth-century sociology. DNE officials such as the black sociologist George E. Haynes drew on emergent theories of Chicago School sociology—specifically, theories of social organization—in delineating black migrants' sudden shift from primary (rural) to secondary (urban) relations. Though the department's predominately black officials characterized their fellow migrants as "maladjusted," they located these deficiencies in social—rather than biological—sources, eschewing previous actuarial assessments of blacks as innately inferior.[7] The department pursued a three-pronged strategy for incorporating migrant black workers into the wartime labor economy: develop a cadre of black labor experts, instill black workers with a new industrial consciousness through worker efficiency campaigns, and use the legislative and mediating mechanisms of the state to forge stronger links between white capital and black labor—bypassing the notoriously exclusionary white craft unions. DNE officials fused sociological theories with practices of racial uplift to create new forms of applied racial expertise. They believed that mastery of this racial expertise could replace coercion with rational self-interest as the impetus for blacks' participation in the wartime labor economy—and modernize black labor writ large.[8]

The Vexing Negro Problem: Race and Prewar Sociology

Sociology was the preeminent social science dedicated to racial inquiry in prewar America. Yet when early sociologists talked of "race," they were referring to the disparate recent immigrants from south, central and eastern Europe—the "beaten men from beaten races"—streaming into the industrial centers of the North.[9] The Negro, when considered at all, represented a static pastoral figure, immune to modern processes of socialization and unknowable through standard empirical social scientific methodologies: a figure which existed beyond the imaginary pale. An analysis of the first five volumes (from 1895 to 1900) of *The American Journal of Sociology* reveals that while the term "race" appears with regularity, there is not a single mention of the "Negro."[10] Within the sociological literature of the time, blackness was conspicuous by its absence. Whereas Progressive era anthropology according to Lee Baker functioned as a preservationist discipline committed to the study of "out of the way peoples," such as Native Americans, sociology focused upon the prospective assimilation of "in the way peoples," such as African Americans and, to a lesser extent, European immigrants.[11] At the turn of the century, sociologists still viewed political economy as an organic outgrowth of racial difference rooted in biology. In 1901, Edward Ross, one of the nation's leading sociologists, confidently declared, "The economic virtues are a *function of race*."[12] Ross was perhaps best known for his theory of "race suicide," which contrasted the unsettling lack of elite fecundity contrasted with the high birth rates of the inferior masses. Whereas Hoffman's actuarial narratives of the "vanishing Negro" implicated blacks in their own demise, Ross's narratives of race suicide linked a myriad of urban and industrial problems to the reproductive practices of white elites.[13]

By the turn of the century, race slowly displaced class as the prime focus of social divisions in America. Sociologists at the University of Chicago posited a model of human progress that detailed the maladjustments and accommodations of disparate peoples through successive stages of social development. Modernity's ability to annihilate space and time uprooted traditional hierarchies and threw populations into perpetual transitions. Sociologists charted these shifts along a rural and urban axis. Primary (*gemeinschaft*, or *rural*) relations were bound by custom and tradition, while secondary (*gesellschaft*, or *urban*) relations were those voluntary and pragmatic behaviors engendered by modern capitalist industrial societies. Scholars such as Ross, Robert Park, and W. I. Thomas blended theory with intensive empirical research to chart how the transition from primary to secondary social relations altered social mores, or "folkways."[14] The theory of folkways, developed by Yale sociologist Charles Sumner, provided a means to trace racial progress in sociological rather than biological terms ostensibly freeing individuals from the repressive strictures of race.[15]

Nevertheless, social and biological theories of racial progress continued to coexist in the minds of contemporary sociologists. In February 1907, the young Chicago School-trained sociologist Robert Park addressed students at Tuskegee Institute on the subject of the "Negro problem." Juxtaposing the racial heritage of African Americans with the Japanese, Park remarked that "while the Japanese had acquired great technical skill in manual labour and had attained such habits of thrift and industry . . . no laborers in the world, save the Chinese, could or can now compete with them, the negro has these two things yet to learn." Moreover, while the Japanese had developed a complex civilization of their own prior to their encounter with the West, "the Negro of course had not found himself in such a condition for he had nothing in particular to throw overboard save his religious superstitions." But Park remained hopeful about blacks' ability to adjust to the rigors of industrial modernity and concluded his lecture on an optimistic note: "Least of all should you or any of your race be discouraged by any facts in regard to your people that investigation and science makes known to you. . . . As soon as the negro finds out where he is supposed to be racially inferior he will rouse himself and prove that he is not. . . . Knowledge is power and when he realizes his weaknesses, knowledge will give him the means to overcome them."[16] Despite his patronizing tone, Park's insistence that knowledge and learned behaviors offered blacks a path to civilization was an implicit rejection of biological models of black inferiority conflated with destiny.

Theories of blacks' apparent adaptability, or plasticity, informed much of the cotemporary sociological discourse. Even Park's seemingly dismissive characterization of the Negro as "the lady among races" bespoke an adaptability on the part of blacks'—albeit one born of an "expressive temperament": blacks' innate vacuity reimagined as a source of their ability to be remade."[17] At the 1908 meeting of the American Sociological Society, Alfred Stone reiterated Park's optimism regarding blacks' progress: "I know of no other race in history [who] possess in equal degree the marvelous power of adaptability to conditions, [in] which the Negro has exhibited through many centuries and in many places."[18] Months later, in the pages of the *American Journal of Sociology*, Charlotte Perkins Gilman wrote, "The negro is developing an ability to enter upon a plane of business life and further admitting, most cheerfully, that this proves the ultimate capacity of the race to do so; however there remains the practical problem of how to accelerate the process." Though Gilman held out hope for reforming the Negro, she remained convinced of blacks' innate inferiority, believing it to be the "very condition of our [whites'] advantage."[19] The tone of the "Negro problem" began to shift as heredity gradually ceded ground to emergent social sciences, such as sociology, which privileged processes over biology. Franklin Giddings, in a direct repudiation of Hoffman, stated, "The negro is plastic. He yields easily to environing influences. Deprived of the support of the stronger races, he relapses into savagery, but kept in contact with whites, he readily takes the external impress of civilization, and there is no reason to hope that he will not acquire a measure of its spirit and life."[20] Yet despite a shift toward sociological and environmentalist models of racial development, the current of Negro pathology remained—a potentially reformable pathology—but a pathology nonetheless.[21]

Across the color line, black sociologists struggled to delineate the contours of the elusive so-called Negro type. Foremost among these was W. E. B. DuBois with works such as *The Negroes of Farmville* (1896) and *The Philadelphia Negro* (1899), and his role in the establishment of the renowned sociology department at Atlanta University. Through these various scholarly and institutional endeavors, DuBois sought to infuse an authentic empirical objectivity into social science and "put the science into sociology through a study of the conditions of [his] own group."[22] Yet despite these pretentions to sociological empiricism, DuBois retained a lingering faith in racial essentialism first outlined in his *Conservation of the Races* (1897), a Hegelian ode to race as a historical imperative born of innate cultural essences. DuBois soon shifted from this essentialist formulation of race—though he never entirely abandoned it—in part because he recognized the impossibility of trying to dismantle essentialist (or heredetarian) models of black degeneracy with more of the same; thus, changing the script was the only option. His production of alternative and sometimes radical methodologies was also a means to distinguish himself from many of his fellow "race experts" across the color line.[23] DuBois's subsequent commitment to empirical inquiry and elite expertise as a means of racial uplift stood in stark contrast to Bookerite advocates of more grassroots inflected black vocational education. DuBois linked the practice of sociological inquiry to the theory that the race would be saved by its best men, creating praxis of racial uplift for a generation of African American sociologists.

Black migration engendered the sociological knowledge required for the creation of a laboring Negro type in early twentieth-century America. Greater occupational opportunities, the insidious effects of the boll weevil on the Southern cotton economy, and a desire to escape the daily indignities and horrors of Jim Crow drove hundreds of thousands blacks northward. In 1900, 10% (880,000) of blacks lived in the North, while 89.7% (7.9 million) resided south of the Mason-Dixon Line. Between 1900 and 1915, just over one hundred-fifty thousand blacks moved north, followed by an additional five hundred thousand more in the years 1916 to 1918. From 1909 to 1919, Chicago's black population increased by a staggering 148% (from 44,000 to nearly 110,000), while New York witnessed an increase of 66% (from roughly 92,000 to 152,000) between the years 1910 and 1920.[24] This migration was almost exclusively urban in character, creating a "divorce from the soil on the part of the colored race."[25] These shifting demographics brought the previously ignored figure of the "American Negro" into the mainstream of sociological investigation. Indeed, the very act of migration on the part of blacks undermined prevailing notions of their innate lassitude and docility. Mass migration to the North meant that, "for the first time in history, the Negro had a chance, to get his hand upon the thing by which men live, to become for the first time a real factor in the world of labor." To replace the dwindling pool of immigrant labor and escape the repressive sharecropping system of the Jim Crow South, African Americans were exhorted by officials of the National Association for the Advancement of Colored People (NAACP) to "Come North," where the "demand for Negro labor was endless." As the "Negro problem" went national, northern social scientists such as the sociologist Howard Odum were forced to engage with "race contacts as present realities, not distant theories."[26] Whereas observers such as Hoffman viewed urbanization as a factor in black degeneracy, sociologists like Franklin Giddings saw it as the best "remedy for the Negro's innate racial handicap of indolence." Giddings believed that the "dynamic association" provided by city living would stimulate blacks to ever-higher levels of civilization and wake them from their premodern slumber.[27]

Applied sociology of the Chicago School sought to reconcile disparate peoples to the rhythms of industrial modernity.[28] One of its most adept practitioners as it related to racial matters was George Haynes. Born to former slaves in Pine Bluff, Arkansas in 1880, Haynes matriculated at Fisk University and Yale University, and became the first African American to be awarded a Ph.D. from Columbia University.[29] His doctoral thesis, entitled *The Negro at Work in New York City* (1912), was guided by three overarching questions: (1) Where were blacks employed?; (2) Were they efficient and successful in said employment?; and (3) How were they viewed by their employers and fellow employees? Haynes sent questionnaires to one hundred New York City employers inquiring about their experience with Negro labor. Employers were asked, "In comparison with white workers, were blacks, faster, equal, or slower in speed, better, equal, or poorer in quality of work done, and more, equally, or less reliable?" The majority of respondents stated that while blacks had generally measured up to whites, the "negroes usually had to be well above the average to secure and hold a place in the skilled trades."[30] Unlike

DuBois in *The Philadelphia Negro*, Haynes did not attempt to forge a theoretical model of sociological inquiry. Instead, Haynes hoped his study would serve as "a small contribution to the end that efforts for betterment of urban conditions of Negroes be founded upon facts," which he hoped to place at the disposal of public and private managerial elites towards the development of more racially equitable hiring practices.[31]

Haynes rejected the prevailing social scientific consensus that migration and urbanization were contributing to African American degeneracy. Rather, he saw the "increasing urbanization of the colored race as an inevitable process" and a sign of racial progress. Both Giddings and Haynes argued that blacks' migration to urban centers was driven by the same factors that drove whites, citing "the divorce of workers from the soil brought on by technological advances which reduced the relative need for agricultural workers; the shift from handicrafts to the factory and the subsequent growth of commercial and industrial centers."[32] Yet many of Haynes's theoretical models failed to incorporate migrants' (often extensive) prior experience of southern industrial labor and urbanization. The aforementioned "divorce from the soil" was belied by figures that showed how, in the case of Chicago from 1915 to 1918, only 25% of migrants to the city had originally been employed as agricultural labors in the South. However, approximately three-quarters of the remainder had between five and ten years of experience in a combination of lumbering, railroading, iron and steel industries, and brought with them an acute understanding of the labor processes and lifestyles of urban industrial modernity before coming north.[33]

Much like white migrants, blacks went north in search of expanded industrial opportunities, better health care and education, and the lure of the city's "varied amusements and the contact of the moving crowds."[34] Urbanization was the Negro's destiny whether in the South or the North, and "any back-to-the-land movement should recognize that along with the whites, Negroes will continue to migrate to the urban centers and that they will come to the cities in comparatively large numbers to stay."[35] Haynes directly challenged Hoffman's thesis of congenitally degenerate urban blacks, noting "the problems which grow out of [blacks'] maladjustment to the new urban environment are solvable by methods similar to those that help other elements of the population." Haynes cited the need for "good housing, pure milk, water, sufficient clothing which adequate wages allow, and street sanitation all of which have their direct effect upon ones health and physique," further proof of the environmental basis of race development.[36]

Throughout the immediate prewar era, black sociologists continued to argue for the liberating potential of the urban political economy. Sociologists such as Haynes based their arguments on a radical reconfiguration of Sumner's aforementioned "folkways" of social development. Haynes acknowledged the importance of folkways in charting the shift from primary (rural) to secondary (urban) social relations, but refused to see them as fixed and necessary customs. Instead, he characterized folkways "as human social forces with trends that may be directed or altered by means of well planned educational methods."[37] The breakdown in social mores that accompanied this shift—and the intensely atomized nature of urban life—benefited blacks by diminishing the policies and practices of white supremacy rooted in various local and regional forms of coercion and social sanctions. Within the vacuum of industrial modernity, black workers could potentially shed the debilitating taint of blackness and its connotations of laziness, ignorance, and inefficiency and work their way to respectability. Haynes drew on Chicago School theories of ethnic cycle to argue that over time, familiarity would eliminate animosities between peoples and foster mutual empathy through organic non-coercive social networks.[38] Moreover, unlike politically suspect European immigrants, whose loyalty had come in for constant wartime scrutiny, black elites such as Washington were quick to note that "Negroes need no Americanization" and could be counted upon to serve their country faithfully.[39] The trope of the loyal, docile Negro was in large part predicated on an enduring assumption of blacks' mental incapacity for political radicalism. As premodern savages, blacks could be stirred to violent unthinking outbursts, but only under the tutelage of more "advanced," "devious" peoples—external agitators such as Jews—could they be spurred to articulate these outbursts in political terms.[40]

Like Booker T. Washington, Haynes maintained an abiding faith in the liberating nature of the free market. Accordingly, Haynes linked community development to questions of worker efficiency and wages. He firmly believed that race prejudice "would wither in the face of reason and it will then follow that our attitudes and ideas of an inferior and superior race will change, and our ideas of we the people will expand to include black and brown as well as white." For Haynes, "folkways" offered a "reassuring framework for understanding group conflict and maladjustments as an inevitable part of urbanity and modernization" as well as a powerful rejoinder to biological explanations and eugenic remedies for racial unrest.[41] Haynes's pioneering studies of black workers in New York City, Chicago, and Pittsburgh posited the Negro worker as a knowable object of inquiry, as opposed to the caricatures of the "happy darkey" or "sinister black deviant" that characterized contemporary- and southern-influenced sociological thought. Moreover, it presaged the bulk of blacks' wartime migration, introducing the figure of the Negro to the northern social scientific imagination prior to the actual physical presence of masses of blacks.[42] Haynes was confident that through the dynamic policies and practices of industrial urbanization, the true contours of the heretofore-elusive Negro type would be thrown into sharp relief.

The Negro Comes North: Sociology, Racial Uplift, and the National Urban League

Applied sociology and racial uplift coalesced in the formation of the National Committee on Urban Conditions Among Negroes (NUCLUN) in the winter of 1912. The *New York Times* described the committee as dedicated to "the integration of Negroes into the urban order." The NUCLUN—which in 1920 became the National Urban League (NUL)—grew out of the "new scientific philanthropy," emanating from the New York School of Philanthropy at Columbia University. Professor Albert Wright of the University of Wisconsin elaborated on the new philanthropy: "On the philosophical side, it studies causes as well as individuals and groups, while on the practical side it tries to improve conditions. Philanthropy must be raised to the rank of a science where the practical and the theoretical are yoked together."[43] The *Times* deemed NUCLUN unique, not only for its focus on the "previously neglected urban Negro," but also for its "frank recognition of the prevalent duplication and overlap of social work" and its use of social scientific methodology in addressing "the growing Negro problem." In its first annual report, the group pledged to "do constructive and preventive social work regarding the economic conditions of urban Negroes; cause cooperation and coordination among existing agencies where necessary; and secure and train Negro social workers."[44] The emergence of the NUCLUN cemented the coalition between black and white progressives, linking the social sciences to racial uplift at the intersection of the private and public spheres.

The outbreak of war in Europe in August 1914 put an end to widespread immigration from the continent and put a premium on industrial labor stateside. Prior labor management policies of pitting one ethnicity against another seemed inadequate given the demands of wartime efficiency and the anxieties surrounding what Theodore Roosevelt maligned as "hyphenated-Americans" from eastern and southern Europe.[45] What followed was an interrelated attempt to consolidate native-born white and "near-white," foreign-born working classes into a singular white American identity, while simultaneously incorporating black migrant labor into the republican body politic along carefully proscribed lines.[46] The NUCLUN, now renamed the National Urban League (NUL), became a key mediator of the first wave of the Great Migration, especially in the greater New York area. Through vocational education, health, housing, and job placement programs, the league tried to integrate black labor into the various mediums of the public and private spheres. Due to financial and organizational difficulties, however, the league's initial efforts to do so were largely unsuccessful. For example, from 1913 to 1914, despite aggressively courting New York's business community, the NUL found work for only 181 of 800 black job seekers, and the next year it placed a mere 308 workers out of 1,557 applicants in private and public sector jobs.[47]

Urban League officials conceded that while a lack of socioeconomic capital dictated that blacks temporarily cede financial control to their white allies, they were adamant that blacks control the league's production of sociological knowledge. To this end, Haynes established a training center for black social workers in the School of Social Work at Fisk University in Atlanta. The training called for a two-year program of coursework, after which graduates were dispatched to various NUL affiliates to develop vocational and efficiency training bureaus. These bureaus were designed to inculcate southern black migrants with an understanding of time-work discipline and provide the requisite mechanical skills and knowledge needed to become efficient industrial laborers. Despite the fact that a majority of migrants were already well acquainted with the time-work discipline of southern agriculture, NUL staff constantly admonished recent migrants to "husband one's private time accordingly in order to fully reap the benefits of their labor."[48] NUL night classes in labor economics outlined the socioeconomic obligations of the wage contract for recent migrants in the hopes of alleviating labor turnover which, according to employers, was caused by blacks' seeming indifference to the temporal demands of wage labor. Notwithstanding the league's constant efforts to divest migrants of their seemingly debilitating premodern heritages, the league consistently organized fresh-air outings for migrant children to escape the congested degradation of the city and experience the physically invigorating virtues of rural life.[49]

Despite its urban character, the NUL enjoyed its greatest success in job placement providing employment to black migrants in the tobacco fields of Connecticut in the spring of 1916. Haynes successfully lobbied the Connecticut Tobacco Company to recruit close to fifty black families and one thousand college students for work, in addition to paying for their transportation from Norfolk to Hartford. By June, almost one hundred permanent workers and close to one thousand student contracts were signed. Before midsummer 1917, there were close to three thousand blacks in Hartford.[50] "Thus was born, right in the heart of Yankee land," exclaimed the *New York News*, a leading black daily, "the first significant move to supplant foreign labor with native labor, a step which has resulted in one of the biggest upheavals in the North incident to the European war, which has already been a boon to the colored American, improving his economic status and putting thousands of dollars into his pockets"—albeit through labor that bore a striking resemblance to the southern plantation labor which most migrants had come north to escape.[51]

The NUL was especially adept in the professionalization of black labor expertise. Speaking to the National Conference of Charities and Corrections, Haynes argued that it was necessary to place social work training programs at black colleges because the best candidates were to be found "among those large groups of select, capable, enthusiastic Negro youth at these colleges . . . [and] because the city conditions among Negroes demand minds and characters which have been molded by a broad course of education." For Haynes, it was essential that the rational expertise of all educated blacks be brought to bear on the "problem of social uplift."[52] Haynes insisted on a college education, field placement, and graduate training for black social workers to prevent their segregation in the wider social work profession. But the NUL faced serious hurdles in its development of black expertise at a time when most black professionals were confined to the ministry and teaching professions— both of which did not require college degrees. Haynes and his colleagues at the league hoped that education at institutions such as the School of Social Work at Fisk University would produce a sustainable body of black social scientific expertise, contribute to the long-term material betterment of the race, and lessen financial and political dependence on white philanthropy.[53]

Rationalizing Wartime Labor and Expertise: Founding the Department of Negro Economics

World War I made the black worker an industrial factor on a national scale. Prior works in industrial ethnography, such as the multi-volume *Pittsburgh Survey* (1909–1914), the massive *Dillingham Commission* (1907–1911), and Edward Ross' opus, *The Old World in the New* (1914) had paid little attention to the black worker, given the limited role of blacks in northern labor markets.[54] In the spring of 1915, the Department of Labor (DOL) created its employment service within the federal immigration bureau to alleviate wartime labor shortages by directing workers into occupations commensurate with their supposed racial capabilities—race being primarily conflated with nationality. Professor Emily G. Balch of Columbia University noted the need for "cooperation of an advanced type implies and requires difference; it involves division of labor in the sense of primary specialization along differing lines," foremost of which was race as understood along a white black axis.[55]

Following America's entry into the war in the spring of 1917, the U.S. Department of Labor was charged with the rapid restructuring of the peacetime workforce onto a wartime footing. In response to northern industrialists' requests for additional labor, the department colluded with railroad companies to bring southern blacks to the north. Northern labor leaders were furious, accusing the federal government of importing scab labor, while southerners denounced the practice as poaching of their local labor supply. In response, the DOL tapped James F. Dillard, a Jeanne and John Slater Fund fellow and acknowledged expert on the "Negro problem," to prepare a study on the national implications of African American migration.[56] Professor Dillard quickly organized a force of five investigators, all southern and all white—with the exception of one black professor from Hampton University. Dillard was viewed as a moderate, an advocate of Washingtonian uplift who had the ear of the New South elite. The committee advised Secretary of Labor William B. Wilson to appoint a permanent official Negro advisor to the department to oversee an affiliated Negro labor agency. While some in the black community argued that a Negro agency further entrenched segregation, many members of the "talented tenth"—the educated elite of the black community—clamored for an appointment to this prospective agency, desirous of acquiring the much-vaunted mantle of federally sanctioned expertise.[57]

The selection process to choose the head of the Department of Negro Economics (DNE) revealed divisions within the black community over the strategic limits of accommodationism and the racial politics of expertise. One of the leading early candidates for the job was Giles Jackson, an attorney and rabid anti-migration advocate who continually enjoined blacks to remain in their ancestral southern homeland. Jackson was a notorious figure within the black community due to allegations of financial irregularities regarding his use of a $250,000 approbation for the Negro exhibit at the 1907 Jamestown tercentenary. Throughout the black press, Jackson was pilloried by the likes of DuBois as a "disreputable scoundrel" and a "disgrace to the race."[58] The editors of The Bee, Washington, D.C.'s leading black daily, wrote, "Giles Jackson has about as much to recommend him for any position, save as dog catcher, as the devil has to recommend himself for a position on the right side of the Lord in Heaven." African Methodist Bishop George C. Clement wrote to President Wilson warning that Jackson's appointment "to any responsible place in government would be regarded by many self respecting Negro leaders as a calamity."[59]

Some critics even accused Jackson of being in the pay of southern capitalists eager to maintain a captive and pliant black workforce. For proof, they pointed to Jackson's The Industrial History of the Negro Race of the United

States (1911), which—despite arguing for African Americans labor fitness—disparaged the idea of migration north as a "pointless endeavor." Perhaps the most damning indictment of Jackson was that he was an amateur who had profited from patronage that heretofore had been blacks' sole means of entry into the federal government. Encouraging the shift from patronage to expertise, the heads of the NAACP, the NUL, and the Jeanne and John Slater Fund pleaded with Secretary of Labor William B. Wilson, insisting that wartime demands required a "Negro expert on Negro problems be appointed to lead the proposed bureau." The three top administrators at the Labor Department—Secretary Wilson, Assistant Secretary Post, and War Labor Administrator Felix Frankfurter—all held deep roots in the reform movement and ultimately proved sympathetic to the demands for a less contentious choice to head the DNE.[60]

The anti-Jackson campaign's privileging of professional expertise changed the expectations for a Negro labor advisor in such a way that only a social scientist could be considered for the post. George Haynes's sociological training, moderate politics, and ties to the broader—and biracial—progressive movement made him an ideal candidate. Haynes enjoyed close relationships with the editorial boards of leading reform journals such as *World's Work, North American Review,* and the *Survey,* which designated Haynes its resident expert on black migration. Haynes examined blacks' migratory process through sociological lenses in articles such as "Negroes Move North," which detailed the many maladjustments experienced by migrants while maintaining an unyielding faith that a "respectable work ethic would overcome any of the racial peculiarities of the colored worker."[61] For Haynes, black migrants' industrial marginalization was due to white racism engendered in part by blacks' inefficiency and ignorance. To break this cycle of marginalization, it was necessary to endow black migrants with a new industrial consciousness that would transform them into responsible citizen workers, and thereby earn the respect of whites.

George Haynes was named director of the DNE in the spring of 1918, becoming the first black administrator in the history of the DOL. His mandate was threefold: produce and advance black labor expertise, develop worker efficiency campaigns to "systematically inculcate in the Negro a sense of hygiene, skill development, and contractual obligation," and forge stronger contacts between white capital and black labor through the Negro Workers Advisory Boards. Mixing Washingtonian uplift with a DuBoisian brand of social activism, Haynes envisioned that the DNE and its "social technicians," also known as social workers, would alleviate migrants' maladjustments by endowing them with a new "industrial consciousness" that stressed pride of achievement in personal and collective forms of thrift. From its inception, however, the DNE was plagued by issues of underfunding, understaffing, and equal measures of indifference and hostility from many people in the DOL who viewed it as little better than a "Negro hobbyhorse."[62] The white southerners who had supported Haynes's candidacy hoped that improvements in the productivity of black workers would result in a more stable workforce without undermining the core tenets of segregation: optimizing blacks' labor productivity while maintaining a white supremacist labor economy. Many in the black community hoped that Haynes's experience at the NUL and his desire for an all-black staff at the DNE would expedite the race's industrial progress. However, an analysis of the DNE's activities reveals limited success in their efforts to balance these competing agendas.

Haynes's first task following his appointment was to staff the new division with a cadre of black professionals. Notwithstanding his commitment to interracial reformism, it had always been his intention to hire and maintain an exclusively black staff to foster the development of black sociological expertise. However, Haynes soon found himself struggling to secure capable black men and women, echoing his earlier struggles at the NUL. Across the color line, there was a strong tendency among young professionals to regard the "feminine" nature of social work as a refuge for those unable to make good in the seemingly more demanding professions of medicine and law. Roger Baldwin of the NUL agreed, claiming that "black social work throughout the country numbers many incompetents who have gone into it for sentimental reasons and the relative lack of competition."[63] To make social work a more attractive career for the "race's most vigorous members," and to enhance its masculine capital as a means to racial uplift, the department adopted a five-point strategy, which included a letter campaign to deans and professors at leading black colleges, wide distribution of pamphlets in said schools, publication of articles in leading reform periodicals like *The Survey*, personal visits to social science courses in black colleges, and the establishment of summer and night school classes in social work at selected colleges and schools.[64]

Through his contacts at the NUL and the Fisk School of Social Work, Haynes enlisted groups such as the National Social Workers Exchange, the YMCA, and various black colleges to aid in the recruitment of qualified black social workers. Haynes couched his appeals to social work in the language of patriotism and race pride. Haynes's long search for a deputy ended with his selection of Karl Philips, a young lawyer and respected "race man" who brought a keen scientific mind to the pressing problems of black labor. Along with external recruiting, Haynes drew some key members of his staff from existing government positions. Two of the DNE's longest serving field agents,

Charles Hall of Ohio and William B. Jennifer of Michigan, were U.S. Bureau of Census employees, on loan from the Department of Commerce.[65] The DNE also helped launch the careers of leading figures in black labor studies, such as J. R. Crossland and Forest Washington. By the summer of 1918, the DNE employed an all-black staff of 134 examiners, seven secretaries, and fifteen state representatives. These black professionals, though working as representatives of the DNE, were attached to separate colored sections of the United States Employment Service (USES), concentrating on problems relating to black wage earners in their respective locales. In day-to-day operations, blacks and whites often worked side by side in USES offices, without incident, under orders from the (white) state secretary.[66]

The DNE afforded dozens of black professionals an unprecedented opportunity to continue their education and develop expertise at the upper levels of the federal government. Each of the department's fifteen state supervisors reported to the United States Employment Service (USES) supervisors in Washington, D.C. Although many white USES officials found the dual reporting frustrating, it illustrated a clear example of how black social scientists were attempting to carve out a niche for black expertise within the various institutions of the wartime state. State supervisors sent Haynes monthly reports detailing the number of black migrants, types of available housing, and employment opportunities. Haynes then sent summaries of their reports to the Secretary of Labor, retaining a second set of copies for use in DNE publications. As a minority agency within the Department of Labor, the DNE constantly struggled to ensure that its knowledge production remained the exclusive property of blacks and was not subsumed into the wider bureaucracies of the wartime state.[67]

Creating the Ideal Negro Worker: Rhetorical Narratives of Racial Fitness and DNE Efficiency Campaigns

The DNE's efforts to educate black migrants in the occupational and social mores of industrial life took place on and off the shop floor. Drawing on previous NUL campaigns, the DNE instituted "efficiency campaigns" to instill a new industrial consciousness in migrants, while assuaging white fears about the inefficiency and unpredictability of black labor. Due to budget constraints, the DNE was unable to launch widespread educational initiatives in black colleges or factories—instead relying on a limited publicity campaign to target such institutions. Haynes used his contacts in the reform press to secure speaking engagements with various industry officials and to publish and distribute material on the virtues of efficiency. A popular pamphlet entitled "Why He Failed" detailed the traits of the dysfunctional worker: "He did not report on time, watched the clock, loafed when the boss was not looking, stayed out all night with the boys, failed to show up on Blue Monday, demanded a holiday on Saturday, and lied when asked for the truth."[68] Through materials such as these, DNE officials linked race pride to patriotism to caution blacks that "every time a Negro falls down on the job, he pulls down his country and the entire race"—a strategy that reaffirmed a wartime commitment to racial and national collectivism.[69]

Agricultural and industrial labor both fell under the purview of the DNE, yet the various efficiency campaigns focused almost exclusively on the latter. Officials did everything in their power to distance blacks from their supposed premodern, rural past. Industries that lent themselves to Taylorist standardization—such as the iron, munitions, meatpacking, and shipbuilding industries—were ideal for demonstrating the mechanical skill and discipline of black workers. Shipbuilding, with its demand for "ships, ships, and more ships," represented the ideal medium of patriotic production through which Haynes worked to reframe the ideal black worker. Haynes noted that there "was no more interesting or important work than that being performed in the wartime shipyards." Massive shipyards like Hog Island, located southwest of Philadelphia along the Delaware River, were seen as marvels of industrial organization. Bainbridge Colby, a trustee of the U.S. Shipping Board, declared Hog Island "the greatest piece of industrial enterprise that any country has ever known or has ever made" and the embodiment of the nascent military industrial nexus.[70]

Sociologists found the diverse workforces at shipyards especially intriguing. At Hog Island, African Americans comprised approximately one-fifth of the island's workforce by the war's end. The records of the U.S. Shipping Board (Emergency Fleet Corporation) estimated that approximately 24,000 black men worked in the nation's shipyards during the war, with the number dipping to 14,000 at the close of the war through to September 1919.[71] Just under half of the blacks employed during wartime (11,991) were stationed in the southern district, with the remaining 12,000 concentrated in the Great Lakes, Delaware River, Mid-Atlantic, and Gulf Districts. During the war, 20.7% of black shipyard workers were engaged in occupations classified as skilled. Following the November 1918 armistice, the percentage of skilled black laborers in the yards rose to 27.47%. The largest number of blacks in skilled occupations, both in steel and wooden ship construction, was in the southern district, revealing the wartime state's reach in incorporating African Americans into modern industrial structures and processes across regional lines. Not only did blacks enter the skilled and semiskilled occupations during the war in large numbers, those that did remained in those occupations in larger proportions than blacks in unskilled occupations.[72]

For many observers, the shipyard worker represented the ideal industrial worker: a perfectly calibrated model of physical efficiency. Yet as Bruce Nelson argues, the skilled and mechanistic nature of shipbuilding stood in stark contrast to the equally demanding but far more rough-hewn and dangerous labor of longshoreman, noting that the "hard physical labor on the docks and at construction sites engendered its own mystique of manhood."[73] The staggering degree of mechanical and manual aptitude required for such work afforded shipyard workers the skilled and resolutely masculine character so desperately sought by black male laborers. At the 1918 Independence Day rally in Raleigh, North Carolina, Haynes lavishly praised Charles Knight, a black riveter from Baltimore who drove a record 4,875 studs in nine hours into the hull of a battleship.[74] Knight, "a highly respectable and industrious citizen" of Baltimore, was acclaimed throughout both the black and white press and was even awarded $125 by the British Government for his efforts. The *New York Times* noted that despite Knight's accomplishments, they did not lead to an increase in the number of blacks employed in shipyards. The *Times* observed, "The same condition of affairs, remained in the employment of Negroes at Hog Island. After they had manifested the same evidences of efficiency, they suffered from most invidious discriminations while endeavoring to contribute to the winning of the war."[75] Knight's perseverance in the face of unrelenting racial prejudice made him an ideal model of the patriotic, efficient Negro type—albeit one that was not representative of a wider acceptance of skilled black labor throughout the wartime labor economy.

The DNE's productive narratives of race pride coalesced in Edward Burwell, a native of Philadelphia. Burwell was the captain of a pile-driving crew, who broke the world's record in driving piles at Hog Island in the summer of 1918. Burwell and his crew drove 220 sixty-five-foot piles in nine hours and five minutes, amid a terrific downpour, smashing the previous record of 165 piles in nine hours.[76] Burwell's employer, The Arthur McMullen Company, had been contracted to drive 21,434 piles, yet Burwell and his crew alone drove about 20% of this number. When asked how he came to break the record, Burwell reportedly smiled and pointed to a placard that read: "If at first you don't succeed, try, try, again." According to Burwell, "The sign filled the crew with enthusiasm. We decided one night that a new world's record would be made on the morrow, and it was. Of course we had our little mechanical troubles, but instead of fretting and fuming the men just glanced at the sign and started in with renewed vigor and the record was smashed." Employed as a pile driver for almost fifteen years, he explained that he had never been on a job as large as this before and that it was due to a rivalry with another Negro foreman that he had made up his mind to go after the record.[77] Burwell's sober work ethic, affability, and mechanical acumen embodied the very model of racial fitness that the DNE was working to create.

The war also disrupted the gender dynamics of prewar labor economies. Black women shared in these opportunities to some extent, especially in the meatpacking and tobacco industries. From December 1918 to June 1919, over three thousand black women labored in the nation's stockyards and abattoirs, while another eight thousand were employed in the preparation of chewing tobacco and snuff along with other low-skilled jobs such as cleaning, curing offal, or stemming. DNE officials identified only 136 women in the stockyards employed in the highly skilled occupations of trimming and a mere two working as timekeepers. Likewise, in the tobacco industry, only two women were cited as serving in the capacity of weights and inspections. Among all black workers, this trend towards low-skilled industrial work prevailed throughout the textile, laundry, and munitions industries.[78]

DNE officials believed that the imperatives of wartime efficiency would eventually alleviate racial and gender inequalities in the workplace. However, DNE-sponsored employer satisfaction surveys revealed high levels of employer anxiety regarding black women workers. In a survey of thirty-four employers across a variety of industries, fourteen rated the work of black women as satisfactory, while an additional three found their work superior to white women. Of the seventeen employers who rated black women's work as unsatisfactory, seven cited irregular attendance as the main cause of dissatisfaction while another seven felt that their output suffered from their "incorrigible slackness."[79] Despite the ambiguous nature of these results, DNE officials were sensitive to the use of terms such as "incorrigible," which demonstrated the persistence of hereditarianism models of black workers' inefficiency. To counter these critiques, Haynes and his staff argued that the social constraints of race and the *innate* handicaps of gender combined to limit the progress of black women workers. Pushing for a greater role for women in the workplace, officials constantly stressed the need for workplace hygiene to protect black women workers from the debilitating effects of the industrial workplace. Black women's role as "daughters and mothers of the race" demanded a privileging of both racial and gender identities. For Haynes, industrial employment for men was a precondition for racial progress, whereas for women it was seen as a necessary evil driven by wartime demands, which if left unchecked, could imperil the race's future health. DNE officials' essentialist model of gendered labor mirrored the era's prevailing patriarchal mindset, while simultaneously revealing black men's need to police a black womanhood perpetually under siege by the wider white society.[80]

Wartime working conditions for black women ran the gamut from intolerable to desirable. The disparity in women's working conditions narrowed in industries that were previously dominated by men. For example, the meatpacking and munitions industries generally afforded a healthier workplace than the supposedly "female" industries of textiles, clothing, and tobacco. DNE officials noted that women, both black and white, working at a northern hosiery mill enjoyed no provision for first aid, despite frequent workplace accidents. Workers labored in cramped and ill-lighted conditions for ten-hour shifts with only a half-hour for lunch. The plant had no lockers or lunchrooms and only a single toilet and sink with no hot water. Despite a near total lack of amenities, workers were expected to keep the hosiery free from soiling and were taxed a few cents for each soiled spot found by the inspector. However, a survey of a neighboring shirt manufacturer employing black women revealed a well-lit, heated, and safe workplace fitted with the latest machinery and amenities.

Many wartime workplaces afforded women access to the most up-to-date vocational education, high wage rates, and the chance to augment their pay through generous piecework bonuses. DNE officials were especially impressed with the well-appointed lockers and clean and ample restrooms in one munitions plant that allowed "colored women to maintain their dignity." In October 1918, black women at a small machine works operator in Ohio informed a DNE official of the "resolutely modern" conditions on the shop floor.[81] Black women worked alongside foreign-born women of various nationalities in clean, light workrooms for eight hours a day at decent wage rates. The plant had adequate washrooms, a lunchroom, and adequate first aid facilities. A training school offered vocational instruction at night with no "special arrangements made to race." Despite this apparent race-neutral mandate, company officials acknowledged the tensions between black workers and white forewomen and in the interests of maintaining labor peace, a black forewoman was introduced in the early winter of that year.[82]

Women across the color line saw clerical work as an ideal and respectable alternative to the debilitating and "unwomanly" environments of the industrial shop floor. The wartime transfer of masses of male workers into industrial work created unprecedented opportunities in clerical work for women. For black women, clerical work afforded a professional respectability to individuals laboring under the double jeopardy of race and gender. DNE officials surveyed sixteen offices across the northeast, in both the public and private sector, and found only a few thousand black women employed as typists, stenographers, bookkeepers, and filing clerks.[83] Reflecting the skilled division of labor along racial lines evident in the industrial sector, black female clerical workers were generally excluded from jobs as packing clerks, special investigators, and switchboard operators. The majority of those employed in private commercial offices or government agencies held only temporary contract positions and were eventually forced out by returning veterans. Much like their male counterparts, black women found themselves "the last hired and the first fired." Despite the DNE's efforts to extol black women clerical workers for performing the vital "organizational work in the fight for democracy, erasing all doubts of their love for country," their diligence failed to inoculate them against the postwar reassertion of a white supremacist male labor economy.[84]

Racial Labor in the Wartime Corporatist State: The Negro Workers' Advisory Committees

The DNE's most significant institutional accomplishment was the work of the Negro Workers' Advisory Committees (NWAC). Despite moderate success in developing a cadre of black professionals and creating new narratives of black labor efficiency, the department proved most effective in forging links between white capital and black labor. The chief mandate of the NWACs was to facilitate black labor to meet the demands of wartime industry and mitigate shop floor tensions between the races. Consequently, advisory committees were integrated as an example of the need for interracial unity between workers.[85] Committees organized and mediated conferences between black and white labor workers and employers across nine states and in a number of localities for the duration of the war. As part of this larger process, DNE officials assessed the needs of the labor market, the availability of black labor, and the local racial dynamics to place the right worker in the right job. On August 1, 1918, the U.S. Employment Service (USES) assumed responsibility for recruiting and allocating labor for war industries employing one hundred or more workers. Following this announcement, Haynes worked to coordinate the efforts of the DNE with the USES for the recruitment and placement of black wage earners throughout the nation. Ideally, the DNE would draw on the legal resources of the USES to establish wage parity in various industries "whenever possible" across regional lines. By 1920, NWACs had been established in 225 counties and cities, along with eleven state committees, and boasted approximately one thousand active members.[86]

Negro Workers Advisory Councils served as conduits between white capital and black labor. Though the DNE leadership was not opposed to unionism in theory, in practice, organized labor—specifically the craft unions of the American Federation of Labor (AFL)—demonstrated an anti-black animus, which was tepid at best, if not downright hostile. A pragmatic assessment of organized labor's racial politics led DNE officials to conclude that the race's industrial future lay in forging productive alliances with white capital. Haynes saw the councils as key in allowing black workers to resist the siren song of solidarity that had historically been deaf to the tones of blackness. However, Haynes's pro-capital stance was met with derision by a new generation of black labor activists who increasingly viewed race relations through the lens of the larger class struggle. In the pages of the progressive organ *The Messenger*, socialists A. Philip Randolph and Chandler Owen attacked Haynes as emblematic of the archaic leadership of "old crowd Negroes."[87] Complacent and effete elites hopelessly enamored with superfluous studies and social programs did much to burnish their pretentions to respectability, while doing little to aid the everyday struggles of the black working class. Perhaps the most damning critique leveled at Haynes and DuBois was the allegedly unscientific (or perhaps worse, irrelevant) nature of their work. Randolph dismissed DuBois's *Philadelphia Negro* as "a heavily padded work, filled with superfluous material, very much like a similar work by Dr. George E. Haynes, *The Negro At Work in New York City*."[88] These attacks struck at the very heart of Haynes's and DuBois's faith in the transformative effects of social scientific methodology and their own self-appointed status as race experts.

The DNE's inability or unwillingness to develop a systematic critique of the racial inequities present in the contemporary labor economy was the result of both philosophical and political considerations. Haynes's accommodationist philosophy posited material advancement as a prerequisite to social justice. Contentious issues such as lynching would have to be sacrificed for the acquisition of intellectual and financial capital that would in turn potentially inure black communities from said injustices in the future. In contrast, the editors of *The Messenger* called for new leadership that had the "manly courage" to work for decent wages and housing, protest against segregated army units, and condemn lynching and disenfranchisement.[89] However, Haynes understood that—due to blacks' limited political capital—issues of social justice had to be viewed in pragmatic rather than absolutist terms. At the executive level, President Woodrow Wilson—a proud Virginian and avowed confederate apologist—had little sympathy for black Americans on either a personal or abstract level.[90] Shortly following his election in 1912, Wilson conferred many high profile cabinet positions and federal appointments to fellow southerners who then embarked on a full-scale segregation of the federal workforce. Black employees—from the Treasury Department to the Post Office—were downgraded, fired, and screened out of future appointments. The few that remained were relegated to labor in strictly segregated workplaces throughout the nation's capital—an outcome of policies and practices that clearly coincided with Wilson's stated belief that "segregation was in the interest of the colored people."[91] DNE officials who did challenge the local practices of white supremacy were handicapped by their role as agents of a federal government whose commitment to racial equality was ambivalent at best.

The institutional limits and ideological tensions of the Negro Workers Advisory Committees, and the DNE in general, were revealed in the racial and labor politics of postwar Florida. In the winter of 1918, Governor Sidney Catts warmly welcomed the DNE. From the beginning, various powerful timber interests expressed opposition to Florida's DNE supervisor, William Armwood, a local civil rights lawyer notorious for his previously unsuccessful efforts to unionize black timber workers in the north of the state. Their suspicions were confirmed when Armwood and the USES state director both refused to yield to the timber industry's demand that they prevent the International Workers of the World (IWW) from organizing black turpentine workers. Eventually, Armwood's insistence that the DNE was legally obligated to remain neutral in labor disputes quickly led the timber executives to slander the agency with the proverbial charges of "negro domination" so beloved of anti-black southerners. From Washington, Haynes assured Armwood of his "unwavering support" and refused to order him to make antiunion statements. However, Haynes's resolute defense of Armwood was less about blacks' right to unionize and more about his support for representatives of black professional expertise that superseded any anxieties over offending local racial sensibilities.[92]

DNE officials found the radicalism of unions such as the IWW distasteful to their moderate sensibilities. Nevertheless, they staunchly defended the right of black workers to unionize. In the face of growing pressure from the timber industry, Governor Catts demanded that Armwood be fired by his "carpetbagger masters" and replaced with a "real Florida cracker." When Haynes arrived to mediate the dispute, Florida's state officials and lumber executives walked out of, rather than enter in to, dialogue with a black man who had been endowed with—albeit limited—federal authority. John Kirby, president of the National Lumber Manufacturers, stated that "while I shall be glad to confer with [Secretary of Labor] Wilson, or [Assistant Secretary of Labor] Post if they wish to see me . . . when it comes to sitting in council with Dr. Haynes, a negro, you will have to excuse me. In the South, we tell negroes what to do; we do not take counsel with them."[93] The Florida press accused the DNE and the federal government of colluding with "wobbly Bolsheviks" and attempting to foist "racial equality" on the seemingly beleaguered local white populace. Unable to withstand public backlash, and despite Assistant Secretary Post's objections, the DOL fired Armwood and pulled all black DNE personal from the South.[94] Increasingly powerless to effect real change for black workers, the DNE redoubled its efforts to develop sustainable black labor expertise and promote positive public relations.

The DNE's loss of its southern staff, coming as it did just before the Department of Labor's biannual interagency review, was a debilitating blow to the already faltering agency. An examination of the department's correspondence from the summer of 1919 to its demise in the fall of 1920 reveals its diminished ability to influence federal policy towards black workers. Consequently, Haynes and his small staff felt the effects of marginalization at both the DOL and the War Department. The combustible mix of increasing black migration, returning veterans, and economic depression led to increased racial tensions throughout the nation. Following clashes between white and black workers in cities such as Norfolk, Philadelphia, and Washington, an official at the War Department declared the "industrial experiment of Negro labor" a failure. Freed from wartime constraints, the inferior black laborer now stalked the imagination of postwar elites as the embodiment of disease, unchecked sexuality, and labor radicalism.[95]

Despite the DNE's best efforts, biological notions of black fitness persisted into the postwar era. War Department officials conceded that while racial labor tensions were in part linked to blacks' social maladjustment, they were also primarily the result of the "innate mechanical failings of the colored worker." When the phosphates and navel stores industries in Florida began their reconversion into peacetime production in the spring of 1919, W. F. Coachmen of the Consolidated Navel Stores Company contacted the DOL and the War Department for aid in replacing black workers temporarily recruited for the war effort. Federal officials noted that, due to "the use of modern manufacturing methods, Mr. Coachmen . . . would prefer white semiskilled labor in place of the Negro labor used previously." Reports indicated that while Coachmen and his peers "have not lost confidence in the Negro, they do not feel certain that the improvement the army and the war have brought about in them is a permanent change." Major W. Reynolds of the War Department echoed these sentiments, declaring, "Clearly the negro has failed to adjust himself to the industrial life and must be returned to his natural occupational environment in the South."[96] Whereas wartime imperatives had briefly necessitated the creation of a plastic, malleable Negro worker, the need to reassert white supremacy in a postwar labor economy required blacks' regression to a primitive state.

Following the November 1918 armistice, southern elites aggressively lobbied the DOL and the War Department to repatriate black labor back to the South to aid the ailing agricultural sector. Tellingly, the DNE was excluded from this process. One southerner lamented that, "having secured high wages in the north, the Negro was reluctant to return south and knew of nothing that could induce him to return."[97] Major W. Reynolds hoped that deteriorating economic conditions in the North would force blacks southward: "The whole question of course is simply that the darkies seem to prefer to work in the North than in the South. As they find it increasingly difficult to get jobs in the North, things won't seem so attractive there and by contrast, the attractiveness of the South may increase." To entice migrant blacks to return to the South, the DOL contemplated instituting a standardized wage scale in previously black-dominated industries "certified by someone that the darkies will trust"—effectively a black *pardrone*, an agent of black labor, who would be beholden to local southern authorities.[98]

The dissolution of wartime corporatist alliances frustrated efforts to order the postwar labor economy along racial lines. Less than three months after the armistice, the DOL enlisted the aid of the War Department to secure reduced railroad fares to move Mexican labor from the Midwest to "points of need in the Pacific northwest." However, railroad officials—now free from the demands of wartime political economy—refused the request and the initiative failed.[99] In April 1919, the DOL achieved some success in transferring two hundred blacks from Philadelphia to the tobacco fields of North Carolina at a reduced rate, but conceded that the majority of their requests to secure similar arrangements for blacks had been swiftly rebuffed. Throughout the war, southern elites had vociferously condemned railroad officials for poaching their local black labor force. Yet in the immediate postwar era, shifts in the labor economy led these same elites to advocate for a reversal of this policy urging the federal government to return African American migrants to their southern agricultural locales. The Immigration Service of the USES had requested reduced fares from the Interstate Railroad Commerce Commission in as early as 1908 and had been rejected. The USES made similar unsuccessful requests in 1910, 1912, 1914, and 1916. On June 21, 1918, the Assistant Secretary of Labor wrote to W. G. McAdoo, Director General of Railroads, to request reduced rates for "specific emergency farm labor," and they were again refused.[100] Without the requisite wartime imperatives, various public and private corporatist initiatives to organize labor along racial and spatial lines often stalled in the postwar era. DNE officials eventually shelved these initiatives due to fears that federal intervention in the southern labor economy, on behalf of white capital and black workers, would further engender the wrath of white workers.

Both the DOL and the DNE found themselves in the untenable position of trying to enforce policies predicated on the civil rights of blacks in locales where the very notion of a black citizen was seen as a contradiction in terms. Haynes acknowledged this process was a "delicate and difficult task." White southerners routinely derided the DNE as "that nigger agency" run by "northern carpetbaggers."[101] Moreover, perceptions of the DNE as an agent of "negro domination" were exacerbated by the department's relative legislative impotence. Notwithstanding appeals to the Fourteenth and Fifteenth Amendments guaranteeing equal citizenship under the law, there were no federal laws prohibiting racial discrimination in wartime industries to enforce in the first place. In the absence of these laws, the DNE essentially became an advisory committee that relied on public initiatives such as efficiency campaigns to garner positive publicity for black labor fitness. Faced with a dearth of legal options, the DNE was forced to pursue an explicitly rhetorical strategy that made it especially vulnerable to accusations of "fostering racial discord" and "agitating for social equality." In fact, DNE officials' radical strategy was their only mode of recourse, even though it fundamentally belied the department's intrinsically moderate mandate.[102]

Following the dispiriting events in Florida, Labor Secretary Wilson invited the representatives of forty-five of the nation's leading civil rights organizations to meet with him in Washington during the summer of 1919. After two days of deliberations, the parties agreed on a set of goals to ensure the economic advancement of black workers. The essential elements of the plan called for increased attention by the DNE to create job openings in skilled jobs in public infrastructure projects, provide federal aid to black businesses, develop a national network of Negro Workers Advisory Committees, and lobby for public and private aid to fund black vocational education. Haynes was ecstatic, confiding to Assistant Labor Secretary Post that the implementation of these proposals "would be a boon to the race." However, limited financial resources, a prevailing hostility to federal authority across party lines, and continuing racial violence in the North tempered Haynes's initial exuberance for the DNE as an agent of social change.[103]

The fate of the DNE was sealed in the fall of 1919 when a DOL Civil Sundry Bill appeared before a Congressional appropriations committee devoid of provisions for the DNE. An unnamed conferee raised a parliamentary point of order and cited funding for the DNE as unconstitutional, claiming that the Secretary of Labor had usurped Congressional powers in creating the DNE—

thereby rendering the matter of appropriations moot. DOL officials protested that they had indeed followed proper constitutional procedure in creating the DNE, citing Section 10 of the Organic Act, which directed the Secretary of Labor to "report to Congress any plans for the coordination of activities, duties, and powers of his office with those of other offices relating to labor."[104] An internal memo revealed that on January 9, 1918, Secretary of Labor William B. Wilson had indeed informed Congress that he intended to appoint a Negro advisor to his department and had called on the Labor Department's solicitor to draft legislation for a "permanent division of negro economics." This legislation was introduced but Congress never acted on the bill, thereby precluding any debate or vote on the issue of subsequent DNE funding. Thus, DNE officials were forced to watch helplessly from the sidelines as the department was dismantled.[105]

Civil Rights leaders were incensed over this duplicity. Mary White Ovington of the NAACP, a close friend of Haynes, accused Congress of "bold racial bias." Eugene K. Jones of the NUL asserted, "The appropriation for this important work was killed in Congress through sheer political chicanery."[106] Haynes and Assistant Secretary of Labor Post refused to give up the fight and drafted legislation to submit to the next session of Congress for the creation of a permanent Negro Labor division. Through the summer and fall of 1919, the USES and Bureau of Labor Statistics absorbed some of the DNE's division personal and programs. In a last ditch effort to save the DNE, Labor Department officials arranged for Haynes to testify as the department's official representative before a second Senate appropriations board hearing. Committee Chairman Francis Warren gave the majority opinion, deploying the rhetoric of color blindness to rationalize the postwar rise in anti-black racism: "You are exactly equal under the law. You are exactly equal, of course, under those appropriations. But as far as we are concerned, there should not be a division between classes of workmen, one against the other . . . we have to look at it with the idea of preserving equality. The same rule applies to both."[107] Haynes rejected this disingenuous appeal to color blindness: "The fact is, Senator, that heretofore the inequality has rested the other way when it has come to matters of industrial opportunity and employment." Postwar appeals to class solidarity and racial equality, especially in light of the increasing consolidation of whiteness, were, in fact, contingent on the continued inequality and marginalization of black labor.[108]

Efforts to create a permanent Negro division within the DOL were unsuccessful, in part due to a resurgent white nativism and a postwar backlash against expanded federal authority. Exhausted from the heady reforms of the Progressive Era and the tumult of war, Americans elected the thoroughly unremarkable Warren G. Harding to the presidency based on his vacuous pledge to restore "normalcy" to the nation. Harding's normalcy was characterized by a resurgent nativism determined to purge the republic of undesirable elements such as Jews, Catholics, African Americans, and political radicals of various stripes. The anti-immigration movement, which culminated in the 1924 Johnson Reed Restriction Act, combined with white fears over increased job competition from southern black migrants. This led to a rapid whitening of the ethnic working class under the Caucasian rubric and a hardening of divisions between white and black labor.[109] Though the new Secretary of Labor, James J. Davis, retained a "colored labor advisor," the DNE was effectively dead by the time of Harding's inauguration in March 1921. Emmett Scott, a Tuskegee official and special advisor to the War Department remarked, "I confess personally a deep sense of disappointment, of poignant pain, that a great country in time of need should promise so much and afterwards perform so little." After the DNE was disbanded, George Haynes moved back to New York City, where he became the executive secretary of the Department of Race Relations of the Federal Council of Churches of Christ in America. In this position, he would continue to deploy sociological analysis of race relations within a broader reform agenda.[110]

The DNE's primary focus on knowledge production and anti-racist sociological methodologies was distilled in *The Negro at Work during the World War and during Reconstruction*, published by the Department of Labor in 1921. From the outset, it had been Haynes' intent to publish two full-length studies on wartime black labor. Budget constraints and limited staff, however, forced him to narrow his expectations and publish a single monograph. *The Negro at Work* sought to answer many of the pressing questions regarding the wartime entry of blacks into northern industry: What particular industries did they enter? Were they unskilled, semiskilled, or skilled? What was their employer's estimate and opinion of their work? How did blacks compare with white workers in the same establishments and at the same jobs, regarding absenteeism, turn over, quality of work produced, and speed in turning out quantity?[111] To answer these questions, Haynes and his co-author Karl Phillips assembled a broad set of data gleaned from existing Labor Department records and subsequent DNE investigations of black labor nationwide. Unfortunately, the authors were unable to include data on black labor on the railroads, in the mines, in agriculture, and in domestic service. Although the original plan of the bulletin called for the inclusion of these groups, and although the department did adopt measures to promote their welfare and efficiency, agencies such as the U.S. Railroad Commission consistently refused the DNE's requests for data on black labor. This lack of institutional cooperation revealed the systemic anti-black racism of the postwar state and the apparent indifference to the black workers' experience. The authors were forced to concede that their sample size was limited and fragmentary, and cautioned readers against drawing any broad generalizations regarding the state of industrial black labor.[112]

Nevertheless, material gleaned from *The Negro at Work* yielded key policy recommendations to the Department of Labor. In a September 1919 report, Haynes advised the DNE be given the interagency mechanism to develop greater cooperation with the Public Health Service, Bureau of War Risk Insurance, and the War Department in reducing racial tensions between veterans. Furthermore, he recommended that the DOL allow the NWACs to continue their work in developing publicity and promotional campaigns committed to fostering the Negro's industrial consciousness and forging stronger ties between black workers and white capital. The report also advised that the Federal Government empower its agencies to provide information to employers regarding black workers. Finally, the report suggested that steps be taken with the appropriate state government departments for the establishment of permanent DNE offices, similar to those already established in North Carolina, Ohio, and Illinois. The Department of Labor failed to act on a single one of the recommendations.[113]

The Negro at Work during the World War and during Reconstruction is a key work of early twentieth-century black labor studies and a model of sociological scholarship. Shortly following its publication, co-author Karl Phillips bragged that his statistical charts were accurate to the third decimal and gave "the clearest picture of the current state of the colored labor in the nation." After the demise of the DNE in 1920 to 1921, Phillips remained at the Department of Labor, where he prepared annual statistical reports on black migration and blacks' occupational status for a number of years. Although he responded to inquiries about black workers, Philips had no executive authority and quickly found himself marginalized within the Department of Labor.[114] Throughout the interwar years, black sociologists such as Charles Johnson and Abram Harris drew heavily on *The Negro at Work* and other DNE research materials for their own investigations on black proletarianization in the urban North.[115] Ira Reid, a protégé of Haynes and one of the era's leading black sociologists, remarked that *The Negro at Work* provided the first systematic analysis of the "Negro's entry onto the landscape of modern industry" and the means through which black labor was made modern.[116]

World War I intensified the struggle for power within the factory, increased labor's ability to impose standards and to resist those of the employer, and reconfigured the connections between labor and race management. The

Taylorist drive to scrutinize the entirety of the labor process, to render every-thing and anything as an object to be ordered, was destabilized by wartime ra-cial dynamics at and beyond the point of production.[117] Progressives wrestled with the problems of standardizing labor tasks to wage systems with a seem-ingly intractable workforce composed of immigrants and southern blacks who seemingly lacked the requisite industrial consciousness for the work of war.[118] Yet while wartime labor imperatives accelerated the integration of the immigrant working classes, blacks were ultimately deemed to be beyond the pale of American industrial modernity. Reformers like George Haynes hoped that fostering black labor expertise, developing positive narratives of black labor fitness, and forging stronger ties between black labor and white capital would inculcate an effective industrial consciousness in blacks and allay fears about the unpredictability and inferiority of black labor. However, due to a lack of financial, social, and political resources, DNE officials were forced to effectively speak the ideal black worker into existence through little more than enthusiastic press releases and ineffectual labor mediation prac-tices. Paradoxically, these symbolic gestures—the privileging of theory over practice—allowed Haynes and his staff to preserve their own (limited) social capital as racial arbiters unsullied by the daily realities of black working lives.

The DNE—much like the NUL and the NAACP—represented the fusion of liberal environmentalism and accommodation that eventually came to define mainstream racial thought in early twentieth-century America. This intellectual shift meant that no longer did black elites have to fight racism "with revisionist [essentialist] racial ideologies of their own; now they had science on their side."[119] Yet perhaps the DNE's greatest failure was its be-lief that white racism stemmed from ignorance or animus rather than prag-matic and often naked self-interest evinced by a "possessive investment in whiteness"—an investment characterized less as a "snarling contempt than a system for protecting the privileges of whites by denying communities of color opportunities for asset accumulation and upward mobility." Therein lies whiteness as a function of early twentieth-century labor economy.[120] DNE officials continually underestimated the socio-political imperatives of a white supremacist labor economy—the autonomy of white labor predicated on debased blackness—and the persistent allure of the "wages of whiteness" to white workers.[121] But there is a danger in viewing race and racial identity simply as products of the social imagination, or in viewing racism merely as "a consequence of bad ideas, something which can be traced to some intel-lectual wrong turn." DNE officials' mistakenly believed that by simply "laying the facts bare," the irrationality of racial prejudice would be exposed and race relations could be remade along rational lines.[122] Haynes and the staff at the

DNE recognized that blacks' socioeconomic marginalization, while rooted in material causes, was primarily defined and reified by social fictions of black degeneracy. In order to transcend these fictions, black workers needed to adopt white patriarchal norms evinced in a sober work ethic and the practice of expertise to be effectively incorporated into the republican body politic.

The managerial quest for standardized efficiency saw all working bodies as *deficient by definition* and relentlessly sought to purge labor processes of the stubborn human element that consistently thwarted their attempts for systematic efficiency. Consequently, the ideal worker remained elusive and perpetually out of reach for DNE officials. Given the "innate abnormalities" of black bodies, the ideal working or soldiering body became white by default. The normative logic of scientific management dictated that an unruly and insidious blackness—the most debilitating form of the troublesome human element—was conspicuous even in the physical absence of black workers on the factory floor or the battlefield.[123] Anxieties regarding black laborers' fitness often preceded their actual presence in northern and national labor economies. Conversely, these same imperatives—the supposedly race-neutral quest for "the one best way"—potentially offered black migrants the chance to shed their "racial uniform," which had constrained them in a premodern southern labor economy. Moreover, standardization also held the promise of freeing black laborers from the anti-black animus, which informed many of the "brains under the foreman's cap" and relegated blacks to the most menial of occupations. As racialist ideology regarding the unsuitability and disposability of black labor continued to inform management policies and practices through the war and into the postwar years, the potentially emancipatory nature of industrial standardization failed to materialize for most blacks. A complex set of changes in northern economics and politics—along with a shifting pattern of racial ideology rooted in an imagined understanding of black bodies—conspired to relegate blacks to the bottom of the industrial economy. Most black workers' encounters with industrial modernity— whether mediated through the auspicious of the DNE or the military—were profoundly contradictory experiences of tenuous short-term advancement, counterposed with long-term marginalization.[124]

Sociological models of blackness informed the meanings given to the kinds of work individuals or groups could or could not do, where they could live, and how one defined a sick or healthy body in early twentieth century America. Within the discourse of industrial evolution, the stigmata of blackness marked bodies as lazy, unreliable, and unskilled—making occupational advancement impossible. In 1923, Philip Brown, the Labor Department's Negro Commissioner of Conciliation, conceded that "the Negro has always been a problem in the scheme of civilization."[125] Brown's claim for the Negro as problem echoed a prevailing sentiment among self-professed race experts who saw race—specifically blackness—as an intractable ghost in the machine of industrial modernity. By acknowledging the racial politics of the day, DNE officials tried to reconcile the contradictory proposition that blacks' adjustment and assimilation into industrial modernity was ultimately dependent on them shedding their identities as *black* workers, and becoming *workers* who happened to be black. This particular sociological form of racial uplift presupposed the "Negro" as a social creation born of successive transitions from primary to secondary contexts, and therefore amenable to salutary environmental forces. The subsequent encounters of African American workers with the wartime state via the policies and practices of the draft or postwar vocational rehabilitation dramatically reshaped the ways in which race, labor, and notions of the fit and unfit labor informed wartime and postwar labor economies and further enhanced the authority of the state through the rubric of expertise.

The wartime state was a key agent of black proletarianization both at home and abroad. W. E. B. DuBois was one of the first social scientists to develop a systematic critique of a transnational capitalism sustained through the production of racial difference: "Labor is cheapest and most helpless and profit is most abundant. This labor is kept cheap and helpless because the white world despises 'darkies.'" DuBois consequently characterized the war as "primarily the jealous and avaricious struggle for the largest share in exploiting the darker races."[126] Stateside, the small prewar state was ill equipped for the challenges of the first truly modern war. Officials who had been forced to cobble together a functioning bureaucracy eagerly sought out designated "social experts" to fill the ranks of the wartime state from across the public and private sectors. An institutional inability to standardize expertise effectively reduced the wartime state to an aggregate collection of individual technocrats. The weakness of the prewar state, the ad-hoc nature of its wartime development, and the incestuous character of the American social and natural sciences combined to give select individuals extraordinary power in developing policies regarding the racial division of wartime labor.[127]

3

Measuring Men for the Work of War

Anthropometry, Race, and the Wartime Draft, 1917–1919

"War is the health of the State."
—Randolph Bourne, *The State* (1918)

In the spring of 1918, Henry S. Berry, a promising young medical student, arrived at the Tuskegee Institute in Alabama for induction into the U.S. Army. Prior to his arrival, Berry had undergone a battery of mental and physical tests that had found him "fit and worthy to bear arms in defense of the United States." At Tuskegee, Berry and his fellow recruits underwent intense military training: "Stripped to the waist in the broiling sun, we would go through our exercises necessary to the development of arms, chest, abdomen, and legs." Despite the protests and groans of many of the men, Berry later wrote, "[I] enjoyed the exercises very much for I knew that such training as this would make a better man physically of me, if not more, and at the same time enable me to stand the strain of active service with greater fortitude." Soon he and his compatriots "began to assume the shape and appearance of soldiers." Berry was eventually attached to the segregated Army Medical Corps in France and remained forever grateful to the army for the "hard muscles that now cover my limbs, the cultivated deep breathing that guards me from the white plague and my ability to walk many miles and enjoy the love of clean, pure, fresh air."[1] Through a regime of intensive physical evaluation and training, the army had remade Henry S. Berry into a new man, a new Negro, for the brave new world of American industrial modernity.

Military service was a midwife to quasi-feudal black peoples pregnant with the possibilities of wage labor. Henry Berry's experience of military service is illustrative of how the shifting and often contradictory categories of race and labor were mediated and rationalized through the body in Progressive Era America. Although the First World War did not create the impetus for state surveillance and discipline of the body, it intensified that surveillance and encouraged the proliferation of various regulatory institutions and practices. Scholars have argued that "the military requirements of modern warfare and industry provided governments with a powerful incentive to intervene in new areas of the economy including the construction of men's bodies."[2] Heretofore, the soldiering body had been viewed as a source of social contagion and vice. From the earliest days of the republic to the Civil War, the conquest of the frontier and the imperial wars of the late nineteenth-century soldiering bodies and military encampments were seen as repositories of moral and physical decay from which the populace had to be protected.[3]

Managerial elites armed with the required expertise in hygiene and social reform now sought to reconstitute the soldiering body via the draft, anthropometry (the measurement of the human form), and subsequent forms of rehabilitation. Military service became not only a key mechanism of Americanization, but of moral, social, and physical regeneration. Historian David Roediger reminds us that "the bodily and mental fitness of those being drafted into the army were objects of intense nationalist concern."[4] War linked the health of the citizenry—specifically, conscripts—to the health of the nation writ large, and in doing so, posited the notion of the republican body politic in stark physiological terms. The policies and practices of wartime anthropometry—and later, vocational rehabilitation—lent a corporeal character to radical critic Randolph Bourne's dictum that war was indeed "the health of the state" on an individual, institutional, and collective level.[5]

Notwithstanding the case of Henry Berry, however, most managerial elites remained skeptical of blacks' ability to transcend their seemingly brutish physicality and become efficient soldier/workers. In addition to prevailing notions of blacks' physiological inferiority, many managerial observers drew on longstanding notions—dating from the early antebellum era—of blacks' inability to work free of coercion, whether military or otherwise. Black recalcitrance was contrasted with the apparent malleability of the various near-white races of eastern and southern Europe. In the spring of 1918, the writer Irvin Cobb observed the men of the 77th "melting pot" Division from New York City—so named because of the disproportionate number of new immigrants in their ranks—following their arrival in France,

> I saw them when they first landed at Camp Upton (NY), furtive, frightened, slow footed, slack-shouldered, underfed, a huddle of unhappy aliens. . . . now three months later . . . the stoop was beginning to come out of their spines, the shamble out of their gait. They had learned to hold their heads up; had learned to look every man in the eye and tell him to go elsewhere, with a capital H.

For reformers and militarists—such as Cobb, Theodore Roosevelt, and General Leonard Wood—who had long agitated for the "invigorating benefits of universal military service," conscription was a key medium of Americanization, essential for forging a new and vigorous American body politic out of the nation's disparate peoples.[6] Alone among the nation's races, blacks were seen as prisoners to a depraved physiology that precluded them from attaining the full social, fiscal, and political benefits of republican manhood.

Social scientists from the Committee on Anthropology (COA) and the National Research Council (NRC) used the science of anthropometry to evaluate the health, shape, and fitness of various racial types through the wartime draft. Ales Hrdlicka, head of physical anthropology at the Smithsonian Institution and a founding member of the COA, initially defined anthropometry as "that part of anthropology in which are studied variations in the human body and all its parts, and particularly the differences of such variations in the races, tribes, families, and other well defined groups of humanity."[7] The voluminous prewar Dillingham Commission 1907–1911—charged with studying the various socioeconomic and physiological effects of immigration—had made extensive use of anthropometry as a prospective tool of social policy. In "Changes in Bodily Form of the Descendants of Immigrants," Franz Boas utilized anthropometric evaluations and taxonomies to argue for the mutability of racial physiologies and the potential benefits of state managed assimilation albeit in limited form.[8] Less than a decade later, Hrdlicka and his staff worked to equate military evaluation of bodies with their industrial classification along clearly delineated racial lines. Although the army IQ tests developed by Robert Yerkes are perhaps the best-known wartime accounting of racial fitness, reformers were also extremely interested in "correlating the physical capabilities of various racial types with their cultural character."[9] Eugenicists and COA officials such as Charles Davenport argued that while the "drafted men's mental qualities and behavior are of importance, of no less obvious importance is his physique."[10] Through institutional mechanisms and practices such as the selective service draft and vocational rehabilitation, the Negro was (re)constituted—in starkly corporeal terms—as a key object of inquiry in early twentieth-century American social sciences and industrial management theory.

Progressive managerial elites conceived of the Negro in deeply pathological terms. In the representational political economy of Western society, only white men were seen to possess the requisite rationality and discipline that marked a respectable citizen worker. Non-white races—and women of *all* colors—were characterized strictly in bodily terms, beholden to their depraved physicality "the pairing of passion and passivity typical of all the savage, backward races."[11] David Roediger argues that the social and financial benefits of "the wages of whiteness" often worked to offset the downward mobility experienced by all workers during periods of stress, such as rapid migration or war. Wartime anthropometry conferred scientific and legislative authority on the division of white man's work from colored work and effectively ensured payment of the wages of whiteness in the Progressive Era political economy.[12] Throughout this period, black workers emerged as mere objects of law and social policy. They were seen as weak, disordered, or, in the lexicon of industrial evolution, as a people out of time, and out of place. Historians note that, during the late nineteenth to early twentieth century, theories of black inferiority "remained remarkably malleable, as different groups of whites sought to advance their own interests and invent new rationales to justify certain traditional patterns of black labor." Eileen Boris and Ava Baron concur, arguing that "while racial categories, like gender, have proven historically flexible, they remain ideologically rigid," thanks in part to race's deeply embodied nature.[13]

Linking war to work, COA officials measured racial types through three key sites: the evaluation of the first million army draft recruits; the evaluation of the "multiracial" workforce of the American International Shipbuilding Association at Hog Island in Philadelphia; and the measurement of approximately 100,000 demobilized men in the summer of 1919. These practices were extensively detailed in *Physical Examination of the First Million Draft Recruits: Methods and Results* (1919), *Defects Found in Drafted Men* (1919), and *Army Anthropology* (1921). The COA's exclusively white membership included many of the leading self professed race "experts" of the day: Dr. Ales Hrdlicka, of the Smithsonian, Dr. Charles Davenport, of the Carnegie Institute and head of the Eugenics Records Office (ERO) at Cold Springs, Dr. Frederick Hoffman, Chief Statistician of the Prudential Life Insurance Company, Dr. E. F. Hooton, a Harvard anthropologist, and the notorious white supremacist Madison Grant, of the American Museum of Natural History—to name but a few.[14] Conspicuously absent were leading cultural anthropologists such as Franz Boas and W. E. B. DuBois who would challenge eugenic racial models favored by the likes of Hrdlicka and Davenport. DuBois, perhaps the foremost black practitioner of anthropometry, had for years recognized its role in perpetuating physiological models of black degeneracy. But he was ultimately stymied in his efforts to develop positive black anthropometrical models, due to a lack of institutional capital.[15] Operating at the nexus of the public and private spheres, the COA's cadre of exclusively white anthropometricists exercised a near monopoly on the development, dissemination, and institutionalization of wartime racial typologies.

The COA's leading member was a retiring biologist named Charles Davenport, the very embodiment of Yankee ingenuity and respectability, and perhaps the era's leading eugenicist. Born into a prominent Puritan family with roots in the abolitionist movement, Davenport was a shy yet intensely ambitious individual. In 1892, he obtained a Ph.D. in biology from Harvard and subsequently secured a position in the school's fledging zoology department. As a member of a new "anti-speculative generation of biologists," he was a pioneer in importing the European science of biometrics—the statistical analysis and quantification of evolution—to America in the early twentieth century. Biometrics formed one of the pillars of early American eugenics. In 1904, Davenport was able to persuade the wealthy Carnegie Institute of Washington—with its ten million dollar endowment—to fund the establishment of a station for the Experimental Study of Evolution in Cold Springs, New York. Six years later, thanks to the largesse of the Harriman railroad fortune, he established the Eugenics Record Office at Cold Springs, cementing his position as America's premier eugenicist.[16] While many questioned Davenport's scientific acumen—Boas was an ardent critic quick to deride the "amateurish" quality of his work, Davenport was nothing if not an accomplished self-promoter, possessed of the true faith of eugenics. Perhaps his greatest strength was his ability to frame both complex biological phenomena in stark social terms and eugenics as little more than an applied science committed to the long-term betterment of society. Davenport's characteristic propriety led him to eschew incendiary commentary in favor of what he believed to be the rational commonsensical truth of eugenics. Yet this faith was soon tested by the war, which Davenport saw as an exercise in racial fratricide that imperiled the long-term survival of the white race(s). Only an acute understanding and management of the conflict's racial dynamics and contours could avert this disaster and pull white civilization back from the abyss.[17]

The selective service system and the policies and practices of anthropometry reified prevailing social hierarchies, which linked regional and national norms of race, class, and citizenship. Wartime anthropometry determined the physical requirements of military service, along with "the physical adaptation of workman to highly organized industrial functions." Industrial imperatives were paramount, given that recruits were disproportionately drawn from the industrial working classes—the majority of whom were semiskilled or unskilled laborers. Enoch Crowder, head of the Selective Service, estimated that industrial workers had a participation rate that was as much as five times higher than that of agricultural workers.[18] COA officials argued, "We should relate our army establishment to society in training for our daily social and economic life, wherein young men may achieve not solely military training but equipment as well for the industrial life ahead of them." This equipment included everything from mechanical skill to time-work discipline. In both the military and industrial spheres, physical fitness and efficiency were measured by the body's resistance to disease and fatigue. In the context of wartime political economy, the value of a laboring body resided in its ability "not to crack" under the withering strain of militarized industry and mechanized warfare.[19] Roger Horowitz notes that military service is a special form of work experience, with its attendant issues of cohesion, division, authority, community, and recreation. Yet he warns against "overly ambitious levels of generalization concerning the effects of military service on workers" and stresses the need to develop analytical tools cognizant of the variegated character of military service and its respective historical contexts.[20]

Anthropometry, much like statistics and sociology, was a tool employed by progressives to develop and maintain racial labor hierarchies amid rapid shifts in the nation's political economy. Within a global economy characterized by a highly migratory labor force, increasing division of work processes, labor radicalism, and war, managerial efforts to link bodily form to function were proving especially troublesome. With the chaos of war and rapid changes in technology, the explanatory power of race as an agent of historical development and social change was increasingly in doubt. Amid these fractious conditions, managerial elites turned to the hereditarianism of eugenics as a way to chart races' past, present, and future levels of fitness. Much of the appeal surrounding eugenics stemmed from its ability to historicize and narrate social change—through race and heredity—in the face of modernity's relentlessly futurist character. Eugenics posited individual and group evolution along a continuum of past, present, and future development and ultimately endowed heredity as an unassailable, inevitable agent of progress. Eugenicists argued that, for too long, medical science and anthropology had erroneously focused on "what man had done, and never what man was." Heredity, not environment, was the deciding factor in individual and racial development.[21] Anthropometry linked the historical imperatives of eugenics to the policies and practices of scientific management in its pursuit of the ideal soldiering body.[22] Army anthropometry created a catalogue of racial taxonomies in which martial and labor fitness were linked not only to character but to seemingly immutable traits like skin color and body shape—effectively turning the working body into an index of its laboring abilities.[23] World War I, as the first modern war, provided unprecedented opportunities to measure the constitution of the color line running through the republican body politic.

The Roots of American Military Anthropometry

Founded in February 1917, the National Research Council's (NRC) Committee on Anthropology (COA) sought to reconcile racial form to labor function through the policies and practices of anthropometry. The COA's anthropometrical policies and practices drew heavily on mid-nineteenth-century forms of racial measurement and labor control. American anthropometry had originated in the Civil War, primarily with Col. Benjamin Gould and his seminal work, *Investigations of the Military and Anthropological Statistics of American Soldiers* (1869). An astronomer by training, Gould was hired by the U.S. Army Sanitary Commission to measure the terrestrial bodies of Union troops and create tables of "normal" body measurements. Gould dutifully measured recruits' arms, legs, feet, heads, trunks, heights, and weights—and quantified lifting ability, lung capacity, pulse, and vision. To do this, he used a variety of instruments, such as the spirometer, a bewildering contraption that measured an individual's lung or respiratory capacity.[24] He then categorized his findings by country of origin, age, degree of education, and most importantly, race. Gould concluded that both "pure blacks" and "mulattos" possessed a smaller pulmonary capacity than whites and disproportionately suffered from flat feet. Perhaps, from an evolutionary perspective, blacks' longer arms—the distance from fingertip to kneecap being shorter than whites—marked the Negro as closer in development to a primitive anthropoid than his white peer.[25] These conclusions only fueled speculation among some in the scientific community that blacks were indeed the missing link in the evolutionary puzzle. Gould's evidence of blacks' apparent physiological inferiority was echoed by Sanford Hunt in the *Anthropological Review* in "The Negro as Soldier" (1869).[26] These works, and others like them, informed a larger corpus of late nineteenth- and turn-of-the-century scholarship across the natural and social sciences that sought to trace blacks' apparent degeneration post-emancipation in starkly physiological terms.[27]

Civil War anthropometry informed subsequent nineteenth-century discourses of race and human development across the transatlantic world. Charles Darwin cited Gould's work in *The Descent of Man* (1871) to illustrate the ways in which physical differences both within and between races could be delineated. Hoffman's aforementioned 1896 treatise on black extinction, *Race Traits*, also drew heavily on Gould's work in its quantification and monetization of racial health. Despite the best efforts of sociologists and anthropologists such as DuBois and Boas to develop antiracist forms of anthropometry, they too found themselves beholden to the prevailing belief that racial forms or bodies—however mutable—were ultimately an expression of racial function. In short, races *were* as races *did*. From a professional and ideological perspective, black social scientists were strongly dissuaded from pursuing anthropological study, as it "had no practical benefit to the needs of the race."[28] In the wake of Civil War anthropometry, "no longer would attitudes of racial inferiority have to employ those prewar measurements and conclusions tainted by proslavery arguments."[29] Scientific objectivism had seemingly displaced crass sentimentality and prejudice in the debates over black inferiority. In a bitter twist of historical irony, the war that led to the destruction of racial slavery produced a body of anthropological research that sustained anti-black prejudice well into the twentieth century.

Anthropometry experienced a renaissance with America's imperial forays into Asia and the Caribbean at the turn of the century. The socioeconomic demands of empire foregrounded the soldiering body as a medium of imperial knowledge, reflecting Americans' shifting place in the world. Yet many "old-stock" white elites worried about the dangers of empire on both individual bodies and the republican body politic as a whole. Watching American regiments leave for Cuba in 1898, Madison Grant could not help but be impressed by "the size and blondness of the men in the ranks as contrasted with the complacent citizen, who from his safe stand on the gutter curb gave his applause to the fighting men and stayed behind to perpetuate his own brunet type."[30] The potential of imperial labor to reinvigorate its agents paradoxically weakened the nation by the enfeebled individuals it left behind. Moreover, the physiological effects of empire abroad were still unclear. During the Spanish American and Filipino American wars, and the subsequent occupation of the Philippine archipelago, U.S. military officials conducted various studies on what effect, if any, these foreign environments had on the health of American troops. After all, it was in these "exotic" environments where it was believed that "the white man lives and works only as a diver lives and works underwater."[31] Troops bound for colonial service abroad underwent anatomical analysis to determine their racial fitness and endurance in allegedly debilitating tropical climates.

Imperial labor was fraught with socioeconomic and physiological contradictions. For many Americans, the desire for empire coexisted with an abiding fear of these new dark and uncharted colonial spaces. The science of climatology, which linked racial evolution to climate, was a response to the anxieties surrounding socioeconomic, racial, and gender status that accompanied the rise of American empire. Americans often turned to corporal metaphors to express this ambivalence, counseling them to "take up the white man's burden" while warning them about the debilitating effects of these new tropical spaces, ruefully termed "the white man's graveyard."[32] In the case of the Philippines, U.S. army medical officials estimated that whites required approximately two years to become acclimated to the archipelago's climate. Officials found that a "moral life, with plenty of hard work" was found "to counteract, in most cases, the so-called demoralizing effects of the Philippine climate." Imperial labor would redeem both its architects and subjects preserving the superiority of the former and setting the latter on the path to civilization.[33] Despite its small size, the prewar army was one of the few national institutions with the resources and material to conduct systematic anthropometric and medical assessments of American bodies linking studies of soldiers' health in the Caribbean and the Philippines to a broader discourse of race, work, and the nation's imperial constitution.[34]

Anthropometry revealed the imperialist urge for order and the need to place the right bodies in the right places both at home and abroad. White elites believed that empire could serve as a balm to the proverbial "Negro problem." In 1900, Nathaniel Shaler, a geographer and the dean of Harvard's Lawrence Scientific School, proposed that "troops required for Federal Service in tropical lands might well be recruited from the abler Negroes," whom, "as a result of their tropical constitution, would endure the tropical climates better than whites."[35] In 1901, Alabama senator John Tyler Morgan proposed the formation of black colonies in the Philippines, reasoning that blacks—as fellow tropical peoples—were best equipped to compete with the islands' indolent natives and could perhaps serve as a vanguard of a broader American colonization of the archipelago. However, Morgan's plans were effectively scuttled when it was revealed that after extended duty in Southeast Asia, whatever immunity black troops possessed was compromised by their lack of "moral stamina"—evidenced by their tendency to fraternize with native women—and the deleterious effects of various tropical diseases such as malaria.[36] Whereas interracial sexual transgressions that occurred in the U.S. punished black men with murderous violence, in the colonial context, sex across or within the color line was generally countenanced when the races in question were deemed inferior. Anthropometric studies of black troops delineated the detrimental effects of socio-sexual diseases manifested in lesions and inflammations that marked the black body as degenerate. Within only a few years, blacks had gone from being regarded as acclimated "children of the tropics" to representatives of the vast native reservoir of disease and degradation.[37]

By the turn of the twentieth century, the military body was increasingly acclaimed in civilian life throughout the Western world. Military definitions of fitness were widely adopted in public life contrasting the "A1," or first class, body versus the "C3," or third class, body. Across industries, the male body became a prominent object of inquiry—a tool used to delineate shifts in the nation's political economy driven by increased migration and technological change. Beginning in the late nineteenth century, transatlantic reformers in the social and natural sciences had turned to the medium of conscription as a means to chart social change along corporeal lines. The experience of the British during the Boer War exacerbated fears of working-class degeneration when the nation's elites were stunned to find that few of the nation's industrial workers were fit for service. The nation "awoke to the fact that its vigor was sapped," as its working classes showed up for battle as a "hooligan, anemic, neurotic, emaciated and degenerate race."[38] For many, it seemed as if the working classes—especially those in heavy industries such as coal and steel production—had degenerated into a debased subset of a wider white race, or perhaps even a depraved race unto themselves.[39]

On both sides of the Atlantic, it was widely acknowledged that both the state and employers had to accept some responsibility for rendering the male physique as means to preserve national health. A decade prior to Private Berry's aforementioned transformation, leading progressives such as the feminist Charlotte Perkins Gilman had made the case for industrial conscription as effective race management:

> Let each sovereign state carefully organize in every county and township an enlisted body of all negroes below a certain grade of citizenship . . . For the whole body of negroes who do not progress, who are not self-supporting, who are degenerating into an increasing percentage of social burdens or actual criminals should be taken hold of by the state. This proposed organization is not enslavement, but enlistment . . . To be drafted to a field of labor that shall benefit his own race and the whole community, need not be considered a wrong to any negro. It should furnish good physical training and as much education as each individual can take.

Gilman echoed prevailing notions of essentialist racial thought citing Robert Park's characterization of blacks as the "lady of the races." In order for her prospective scheme of industrial conscription to work, Gilman noted: "The whole system should involve the fullest understanding of the special characteristics of the negro; should be full of light and color; of rhythm and music; of careful organization and honorable recognition. The new army should have its uniforms, its decorations, its titles, its careful system of grading, its music and banners and impressive ceremonies."[40] Yet like all good progressives, Gilman believed that only the state—through its marshaling of expertise and legislative coercion—was capable of providing the so-called inferior races with the requisite tools needed to work their way to civilization.

The outbreak of war in Europe in the fall of 1914 inextricably tied the health of the worker-soldier to the body politic. The wartime extension of workplace safety legislation and public and military hygiene programs revealed the corporatist drive to view national productivity as an organic whole, in which the health of its constituent parts had to be maintained at all costs. However, Americans were hesitant to draw definitive racial conclusions from European anthropometry. Hoffman noted that while "every authority on anthropology on both sides of the Atlantic conceded the supreme importance of race as an underlying determining condition in the physical proportions or dimensions of workers and recruits," there was too much ambiguity regarding "race" in European anthropometry to draw direct comparisons with race relations in the U.S.[41] Whereas Americans linked race to distinct physical markers such as skin color, hair, and body shape, European military anthropometry tended to privilege "intra-racial" rather than "interracial" difference, to privilege place over blood. Evaluating the 1914 to 1915 rejection statistics of European armies, Hoffman concluded that, "the term race is not one which permits of precise definition, for entirely pure races are certainly no longer met with in European countries." Europeans believed an individual's race roughly corresponded with their residency or birthplace: a Prussian was generally accepted as someone who was born in Prussia. Though it should be noted that these spatial "birth place" definitions of race were largely confined to academic and elite anthropometrical debates and did not characterize European racial discourse as a whole. Much like its American variant, European anthropometry understood race as an immutable biological and cultural identity.

Nevertheless, Hoffman found this methodology to be maddeningly imprecise from an American ethnological perspective, arguing for the need to "[correlate] the physical data to the place of birth."[42] Racial physiologies had to be conceived of as a long-term product of both space and time. Hoffman drew on his training as an actuary to note that "the same definite relationship between disease predisposition and inherited ancestral traits has shown to be the case in the inheritance of physical proportions of the body." Hoffman proposed that the U.S. Army abandon its current European-informed models of racial classification—which only required examiners to list recruits' color and nationality—in favor of methods more attuned to models of physical anthropology, which more effectively linked race to place.[43]

The war simultaneously shattered and reaffirmed progressive's faith in rational inquiry. The "war to end all wars" destroyed progressive notions of perpetual progress while simultaneously reconfiguring new forms of pragmatism. Indeed, the war convinced many progressives that there was perhaps nothing wrong with pragmatism—as both an ideological and instrumentalist social philosophy—that could not be remedied through pragmatic solutions. Progressives sought to remedy these seeming contradictions of pessimistic rationality in their efforts to regulate and potentially remake laboring bodies of color viewed as recalcitrant. One observer described this process as "an increased interest in physical measurements and the means of improving them when they are below par."[44] Anthropometry could also alleviate the war's potentially dysgenic effects. War "impoverished the breed" the nation's germ plasma, as the strongest were killed at the front leaving behind the unfit to breed.[45] Dr. T. J. Downing, writing in the *New York Medical Journal*, warned that the danger of postwar racial degeneracy lay in "the possibility that through religion and commerce, the idealism of universal democracy, worldwide socialism and the practical annihilation of distance, the long headed races of Western Europe and America may invite or permit a migration of the mixed or broad head skull types which would be followed by centuries of retrogression."[46] By marking the stigmata of degeneracy in various racial types, anthropometry helped weed out unfit bodies from the body politic. For many progressives, an "accounting of national health" through a process like conscription was vital for addressing questions of "national and racial vitality, physical progress, or deterioration at a time of great unrest."[47] And only the state as the facilitator of racial and labor management expertise could effectively serve as the primary arbiter of national health.

Taking the Measure of the Nation's Racial Constitution: Anthropometry and the Draft

The wartime draft was a massive undertaking unprecedented in U.S. history. Congress formally declared war on Germany on April 6, 1917, and in the following month, conscription was instituted with the passage of the Selective Service Act. The draft soon became a litmus test for progressive principles, and the utility and limits of state coercion as an agent of social change. Over the course of the war, America transformed its small peacetime volunteer army of 126,000 men into a fighting force of approximately 4 million (350,000 of whom were black).[48] Officials drew heavily on the experiences of Civil War conscription such as the infamous racially contentious New York Draft Riots of 1863 eager to avoid past mistakes. Conscription was couched in the rhetoric of a voluntary "selective service," owing to the need to reconcile civil liberty and national efficiency in a modern, multiracial democracy ostensibly committed to a war to make the world safe for democracy. General Enoch Crowder, director of the Selective Service, stated that "conscription in America was not the drafting of the unwilling . . . the citizens themselves had willingly come forward and pledged their service." President Wilson described the process not as a draft per se, but as a "selection from a nation which has volunteered in mass." For the philosopher John Dewey, the irony of the state effectively "press-ganging" men into the armed services to "make the world safe for democracy" was a perversion of American ideals of social justice. Despite fears that June 5, 1917, registration day, would be met with violence, or at the very least indifference, the day largely passed without incident. At the end of the day, 9.6 million men—including 700,000 African American men—had been registered for military service.[49]

Wartime imperatives and long term strategies of race betterment necessitated a clear and full accounting of the nation's physical health, which only the draft could provide. Like most progressive initiatives, the draft was a curious mix of the pragmatic and the utopian. President Wilson, who framed the "war to end all wars" in idealistic terms, continually stressed the practical character of conscription. After signing the Selective Service Act into law on May 18, 1917, he announced, "This is not the time for any action not calculated to contribute to the immediate success of the war. The business now in hand is undramatic, practical and of scientific definiteness and precision."[50] Ales Hrdlicka demanded that conscription "should yield nothing less than the initiation of a National Anthropometric Survey . . . the existence of nations in the future will depend largely on the conservation of the physical standards and soundness of their people."[51] Reformers and military officials linked the regenerative potential of conscription to notions of sexual purity and manly self-control traits generally seen to be the sole purview of white men. Conversely, those peoples lacking such traits were marked as deficient men, workers, soldiers, and Americans.

Unprecedented levels of federal manpower were needed to process the masses of conscripted men. Out of approximately 9.6 million males—ages 21 to 30 years old—who registered for the initial selective service draft in June 1917, some 2.5 million, were examined at local draft boards. After examination at the local level, qualified men were dispersed to one of the nation's sixteen mobilization camps, where they underwent a second or possibly third battery of physical tests. These tests were administered by officers of the Medical Department of the Army under the supervision of the division or camp surgeon. The chief medical officer was responsible for selecting the examining staff, which was drawn in part from regimental surgeons and base hospital personal.[52] Though local draft boards and camp physicians exercised some latitude in interpreting the examination standards—detailed in the Selective Service Act of June 1917—they generally conformed to a common pattern. Regulations proscribed a minimum height of 61 inches, weight of 118 pounds, and a chest circumference of 31 inches for recruits. Physical defects given as causes for rejection included those found in the skin, head, spine, ears, eyes, mouth, neck, chest, abdomen, anus, genitals, hands, and the lower extremities. Draft officials toyed with the idea of expanding the draft age from 31 to 40 years to expand the number of defects that could disqualify one from military service. However, they were less willing to compromise on height, weight, and chest circumference standards. Despite official efforts to standardize fitness, these norms often proved negotiable for "healthy chaps" able to draw on their respective gendered or racial privileges in pleading their cases to the appropriate military authorities. Case in point was a thirty-one-year-old white recruit from Brooklyn, New York—a post-graduate student at NYU and a former athletic director of playgrounds—who, after being placed in Class 1A, was informed by his local draft board, "that new army regulations forbade [his] induction into service because [he] was only sixty-two and a half inches"—a half inch below the required sixty-three inches. The recruit bemoaned the fact that, "as there are plenty of sickly fellows ready to do clerical duty, why is it necessary to pick upon a healthy chap, fit for the real kind of war work?"[53]

The African American intelligentsia reacted to the draft with ambivalence. W. E. B. DuBois hoped that military service would confer full citizenship on blacks. He explained his reasoning in a July 1918 editorial in *The Crisis* entitled "Close Ranks": "Let us while the war lasts . . . close our ranks shoulder to shoulder with our own white fellow citizens and the allied nations that are fighting for democracy."[54] Driven by a mixture of professional and political motivations—not the least of which was a prospective captaincy in the army—DuBois's call to arms was met with scorn by many blacks. Radicals such as A. Philip Randolph and Hubert Harrison argued that black support for the war represented a "surrender of life, liberty, and manhood." Randolph declared that he would fight not to make the world safe for democracy but to "make Georgia safe for the Negro."[55] By the end of the war, draft boards had registered 24.2 million men "acceptable for military service," with blacks comprising roughly 2.3 million, or 9.63%, of the total registration. Across the color line, a significant minority of men resisted conscription by feigning illness, self-mutilation, or simply failing to register. However, it is near impossible to determine whether this resistance was driven by political or religious ideologies or sheer indifference. Roughly one in ten of all American men eligible for registration, and approximately 30% of those selected by the draft boards, refused to present themselves for formal induction into the army. Local racial politics, high levels of illiteracy and—notwithstanding the Great Migration—the overwhelmingly rural character of the mass of black recruits made it difficult for officials to gauge the exact number of blacks selected for the selective service. Although, the available evidence reveals that the majority of black men complied with wartime conscription.[56]

Black support for the war effort stretched across class lines just as it did in white communities. For members of the black elite, the call to the colors was a duty incumbent upon them as the self appointed leaders of the race. Steeped in traditions of military service from the revolution through the Civil War blacks of all classes saw military service as a path, albeit a historically circuitous one, to full citizenship.[57] Of the 750,000 men in the Regular Army and the National Guard at the beginning of the war, approximately 20,000 were black. Nevertheless, longstanding theories of blacks' lack of military prowess—informed in part by Civil War anthropometry—had denied most African Americans the chance to prove their martial masculinity on the battlefield. The reasons were numerous, and included white southerners' anxieties about armed blacks, along with a prevailing sense that the Negro was naturally "yellow" and unfit for the demands of modern mechanized warfare. One American war correspondent remarked, "One who sees the Negro stevedores work notes with what rapidity and cheerfulness they work and what a very important cog they are in the war machinery."[58] Characterizations of black laborers as both childlike "happy darkies" of a bygone era versus vital cogs in the machinery of the wartime state revealed the contradictory characterizations of black labor as an atavistic, juvenile yet mechanistic agent of the wartime labor economy. Remaking the black worker as primitive efficiency embodied led to more than 89% of black recruits being consigned to labor and supply battalions to provide logistical support to the American Expeditionary Force (AEF) in France. Black draftees who did see combat duty served in National Guard Units and regular army detachments units. The most famous black combat regiment was the 369th Harlem Hell Fighters, who served as an attachment of the French Fourth Army and became one of the most highly decorated allied units of the war.[59]

America's polyglot army provided ample opportunity to develop an intricate schedule of racial typologies. COA officials noted that men of "smaller races" (fewer than 60 inches) were unable to carry the required military equipment whereas men over 78 inches, "drawn primarily from the Nordic type," were more apt to suffer from circulatory diseases. Body size also related to the standard army food ration. Troops in a camp containing many "small southern Italians and Jews" required less calories than one composed mainly of "hardy Scandinavians."[60] One AEF officer concluded that the fighting ability of American troops was in direct proportion to the percentage of "old-stock Americans" in various units, claiming, "get a draft outfit filled with men swept up from the east side [of New York with its high concentration of recent southern and eastern European immigrants] and it is just about as unsafe as anything in the army."[61] While many progressives viewed military service as a key mechanism of Americanization for the "near white" races of southern and eastern Europe, anthropometry consistently reaffirmed blacks physiological *lack* of martial fitness continually held hostage to their depraved physiology. Army officials cited draft evaluations to argue that "the poorer class of backward Negro has not the mental or physical stamina or moral sturdiness to put him in the line against opposing German troops of high average education and training."[62] Only 38,000 blacks served in overseas combat units, constituting a mere 3% of the army's combat forces. The vast majority of black recruits were relegated to segregated service battalions as stevedores, cooks, gravediggers and menial laborers working under conditions not unlike those experienced by many blacks stateside in the convict labor gangs of the Deep South.[63]

American military officials believed that efficient standardization—whether in war or industry—necessitated racial differentiation. Determining whether the Negro could, or should, fight, was based on defining just what constituted a "Negro" in physiological terms. Indeed, one of the committee's stated goals was "to assist division surgeons with basic racial problems such as: Is this person to be classified as a white or a Negro?"[64] Consequently officials came to understand race in mechanistic terms as a collection of interchangeable parts which taken together produced the sum of a racial type. Army officials sought to apply these mechanistic models of labor to increase military efficiency in new, and often intriguing, ways. In late 1917, COA officials tried to incorporate famed French anthropologist L. Manouvrier of the École Pratique des Hautes Études, work on marching. Manouvrier argued that the arrangement of men in each battalion section according to the length of their leg, as opposed to their height, would facilitate more efficient marching techniques. Consequently, any potential gaps in the line could be filled instantaneously by men of any height without requiring any redistribution or exchange of "men of various racial statures," thereby preventing the insidious scourge of fatigue so dreaded by industrialists and militarists. However, an initial lack of funding and the abrupt conclusion of the war in November 1918 precluded COA officials from implementing Manouvier's plan.[65]

Increasingly gruesome modes of warfare highlighted previously neglected aspects of a soldier's physiology. Facial proportions were analyzed to fit men for new equipment, like gas masks, and respiratory fitness was measured to gauge specific groups' susceptibility to gas. As a medic attached to the 92nd Division in western France, Henry Berry recalled one such order whereby "'Negro soldiers will be used to handle the mustard gas cases because the Negro is less susceptible than the white.' Why is the Negro less susceptible than the whites? No one can answer!" COA officials argued that a knowledge of "racial characteristics was necessary to decide on the classification when military organizations are being formed on racial lines, such as 'Negro regiments' or 'Slavic legions'" and foregrounding racial labor knowledge as a function of military efficiency amid rapid technological change.[66]

The impetus to measure the exact dimensions of Negro racial types was colored by the socioeconomic realities of Jim Crow. After the East St. Louis Race Riot of May 1917 and the Houston Riot in August of that year—in which the all-black 24th infantry battalion responded to months of race baiting and violence by local whites by killing over a dozen white civilians—the War Department delayed the general mobilization of blacks until mid October 1917 to ease racial tensions. Once blacks began to be called, southern draft boards habitually refused to grant exemptions for even the most debilitating physical ailments. One of the more egregious examples occurred in Fulton County, Georgia where the local draft board discharged 44% of white registrants on physical grounds while exempting only 3% of black registrants. Nationwide, black draftees received proportionality fewer deferrals than whites. Despite constituting only 10% of the population, blacks made up approximately 13% of those serving.[67]

Historically barred from the skilled trades, blacks were hard pressed to claim deferments for "industrial necessity" and were instead shuttled into low skilled labor. Moreover, many black men were quite simply too poor to claim the usual exemptions for husbands and fathers. In a cruel twist of irony, black men's merge army pay and compulsory family allotment—up to fifty dollars a month to an enlisted man's family—"would actually *increase* many a black family's income, wiping out any claimed deferment on grounds of economic dependency." To preserve the segregated labor economy of Jim Crow, local officials also made a practice of inducting those blacks who owned their own farms and had their own families to support, while exempting young, single men who worked for large planters. This, in turn, effectively reified blacks' subordinate position in regional and national political economies.[68]

COA officials argued that race-based anthropometrical investigations of recruits were vital to avoid errors in "the selection of recruits for [war] work involving a wide range of physical aptitude, stress, strain and specific liabilities to disease." However, the Selective Service failed to provide gradations in assigning rejections simply classifying recruits as fit or unfit for military service.[69] By the summer of 1917, COA officials were convinced that army recruiting officials were still totally "indifferent to the racial aspects" of their work. Officials estimated that thousands of black recruits had been rejected for physical traits such as flat feet, which, although "inherent in the Negro, had no bearing on his military efficiency." Conversely thousands more had been deemed fit for active duty despite suffering from the apparently "hidden wounds of colored diseases," such as tuberculosis (TB) and venereal disease (VD).[70] Military officials saw these pathologies "as endemic to the colored race prior to enlistment" and "readily detectable by the trained medical professional or racial anthropologist."[71] Eager to draw upon a captive black labor force military officials sought to strengthen the selection process through the implementation of professional expertise.

In November 1917, Frederick Hoffman wrote to the Army Surgeon General and the Chief of Staff, urgently recommending the "authorization of the scientific re-measurement of selected groups of men of the new national army . . . for the purpose of securing trustworthy racial data strictly comparable with statistics secured during the Civil War indispensable to all investigations into national physical progress or deterioration." Hoffman argued that "if the measurements being made are ignorant of these pathological facts, it is self evident that one of the greatest opportunities for securing such information will be lost."[72] In January 1918, Dr. George Hale, Chairman of the NRC, wrote to the Army Surgeon General, William Gorgas, to reiterate the need for anthropologists attuned to racial imperatives: "I regard this matter as of the utmost importance, both from the standpoint of pure anthropology and its part in American history . . . for various reasons, much time has been lost and I cannot here impress upon you too strongly the great desirability of prompt action." In response, Hale tapped Davenport, now head of the ERO at Cold Spring Harbor Laboratory in New York to oversee the work.[73]

Beginning in December 1917, in an effort to enhance the scientific and medical rigor of the selection process, the Selective Service allowed local boards to appoint additional examining physicians to assist in all future physical evaluations. The COA recommended two separate schedules: one for use in the local draft boards throughout the country, and the other, which contained detailed anthropometric examinations, for use at the cantonments. Schedule A—for use at local draft boards—was a general schedule that included references to the registrants' town, state, county, and draft number. Examiners were instructed to note an individual's color, "white, black, yellow and so on," as opposed to his race. Applicants were required to note their weight, height, and chest circumference and the birthplace of both their parents and grandparents. A second schedule for special examination at the cantonments called for more detailed physiological investigations referencing "the form of the head, its length, and breadth"—measurements deemed to be of "great anthropological importance." The COA was especially keen to chart other racial traits like "the absence of body hair in certain Negro races, the Chinese, etc.," and the variation between the races' respective wing spans. Both schedules provided space for any additional remarks related to "special bodily peculiarities that the individual may show."[74] Secondary measurements echoed those of earlier racial sciences—specifically the notorious practice of skull measurement known as phrenology—in linking variations in body type to skin color and racial identity, a rejoinder to the coming culturist trend advocated by the likes of Boas.[75]

Anxieties regarding draft officials' apparent inability to reconcile form and function coalesced around the phenomena of an apparent rise in "malingering" within the ranks. Akin to "soldiering" in industrial labor, malingering, or shirking, was the intentional withholding or slowing down of the labor process and was seen by those in the military and industrial spheres as a major impediment to efficiency and a direct threat to managerial control. Military officials divided malingerers into two groups: those who "do so [malinger] with full knowledge, intent and responsibility; and those whose members are hypochondriacs or constitutionally inferior individuals."[76] While the first group was relatively small, it was the second group, "the constitutionally inferior," who posed the greatest threat. Reformers feared that the hyper regimented and social nature of military life, along with a basic fear of combat, greatly increased the chance for "systematic malingering, which like its industrial cousin, stemmed not from a man's natural inclination to loaf but from one's self interest in relation to that of other men." One surgeon remarked, "In all of the cases in which defects were simulated or exaggerated, the patients were actually defective either in a lesser degree or in a different affection or both." Military officials were forced to walk a fine line between facilitating cooperation between disparate types of men and militating against any tendency on their collective part to restrict or withhold their labor at the front or behind the lines.[77]

The conflation of malingering with soldiering clearly revealed the increasing militarization of labor. Moreover, the logic of industrial evolution persistently characterized malingering in gendered terms through the rhetoric of dependence that was seen as proof of a congenital lack of manliness. Blacks, as members of a so-called "tropical race," were seen as genetically predisposed to indolence. For whites guilty of such practices, deviant behavior, not genetics, was to blame—though charges of malingering were disproportionately leveled at colored recruits.[78] Arthur Little, a white officer in the 15th New York colored regiment, recorded his reactions to the sight of his men bathing naked in a pond, claiming that it "caused a number of us to exclaim: 'with Henry M. Stanley in Darkest Africa!'" Like most white officers assigned to black units, Little saw blacks as a people governed by unreasoning emotion and brute strength who, if left to their own devices, would surely succumb to state of dissipation.[79] Little saw his "colored charges" as only one step removed from the jungle primeval and the black soldiering body as indolence embodied.

Malingering was a socio-medical condition that was also seen as highly performative. Despite suspected malingerers' various efforts to disguise their condition through "excessively self-assertive" or "overconfident" behavior, military officials felt that advances in physical anthropology and a keen analytical eye for symptoms such as dilated pupils or rapid breathing that were often "indicative of various racial traits" exposed the subterfuge of malingering. Military officials believed that "there was something indefinable yet pathological in the bearing of the malinger which medical experience alone can detect." Though examining personnel were advised to use some degree of "personal discretion" in evaluating suspected malingerers, they increasingly relied on proscribed medical models of physical anthropology that marked bodies as unambiguously fit or unfit. The "Negro type" and the "malingering type" were increasingly seen as one and the same: deviant bodies that had to be identified and weeded from the ranks to preserve military vigor and national vitality.[80]

Ultimately, time and its contractual dynamics were at the nexus of debates regarding malingering and how best to manage black military manpower. Yet endowing blacks with a regimented clock consciousness was merely a necessary precursor to imparting a sense of military labor's contractual obligations and value. But if proletarianization was defined in part by a shift from bonded to contract labor, this transition was complicated by military conscription—for African American serviceman who were only a little more than two generations removed from slavery, the complications were infinite. Building on Peter Way's formulation of military labor as a "marchland of labor relations . . . straddling the worlds of free and unfree labor" provides insight into the curious contractual condition of black serviceman.[81] However, martial wage labor was defined less by its contractual obligations as opposed to those occupations rooted in service and citizenship. This proved problematic for African Americans whose civil rights had been under constant attack by state and federal officials since the dying days of Reconstruction.

Prevailing notions of contract or wage labor in both the public and private spheres were predicated on the autonomy of both the employer and employee: a reciprocity seen to be the sole purview of white manhood.[82] However, with conscripted military labor, an individual's autonomy was ostensibly based on his ability to temporarily cede ownership of both his body and labor to the service of the state. President Wilson's aforementioned description of conscription "not as a draft per se, but as a selection from a nation, which has volunteered in mass" revealed the complicated contractual nature of this process.[83] Conscription relegated partially emancipated blacks to their supposedly "natural" state of bondage by asserting direct ownership over the black soldiering body—albeit only for the length of their military service. Despite receiving a wage, a soldier's productive capacity was ultimately measured by his ability to take life, and if necessary, give up his own in the service of the state. By doing so, he was entitled to various rewards, such as veterans' pensions. Indeed, only a citizen could be subjected to the seemingly contradictory process of "voluntary conscription," because only a citizen was seen to possess inviolable rights that could be temporarily suspended in the service of the state. Black serviceman, seen as congenitally dependent and defective agents, remained beyond the pale of martial citizenship.

In the wake of fears over a supposed spike in cases of malingering, and following the institution of the new medical advisory boards in the winter of 1917, all previously granted exemptions were repealed and subjected to further review. Washington established 1,319 medical advisory boards to allow individuals to appeal the decision of their local draft boards.[84] Selective Service regulations regarding physical examinations by local boards henceforth "prescribed standards of unconditional acceptance and rejection." All cases found upon examination by a Local Board to fall "between these two standards shall be referred to the Medical Advisory Board." The functions of the Medical Advisory Boards, outlined in Section 44 of the Selective Service Act, were to "examine registrants sent to them by Local Boards or State Adjutants General for examination and advise such Local Boards concerning the physical condition of such registrants." The power to determine whether an individual was either accepted or rejected for service, however, remained with the local draft boards. Under these new guidelines, all past, present, and future draftees were required to fill out questionnaires concerning their industrial status and exemption eligibility.[85]

The revised guidelines of the Selected Service sought to establish clearly defined hierarchies of military efficiency. All registrants were reclassified into five classes, "in inverse order of their importance to the economic interests of the Nation." These categories were implemented at both the local and regional level. Most men fell under Class I, defined as "unmarried and married men whose wife and children are not dependent on their earnings." Classes II through IV included men who were the sole support for their dependents, as well as clerics and criminals, whereas Class V consisted of those "totally and permanently physically and or mentally unfit for military service." Those defined as Class I—liable for immediate military service—were dispatched to one of sixteen national mobilization camps to potentially undergo a second battery of tests to determine whether they were fit for military service. The relative impotence of the Medical Advisory Boards deepened with the growing expectation that no individuals "found by the local boards to be qualified for military service will be rejected upon their subsequent examination by any examining surgeons at the camps or cantonments."[86] To the chagrin of many COA officials, military expediency continued to trump long-term considerations of race betterment and national health.[87]

Anxieties regarding the nation's seemingly enfeebled manhood were ultimately rooted in competing claims of expertise, a characteristic of many Progressive Era debates. Critics noted that draft boards were made up of three local residents of the respective county or city, only one of whom was a physician—although often with dubious professional accreditations and beholden to the dictates of local politics rather than long-term race betterment. Davenport and COA officials blamed the high rejection rates on local draft boards' misunderstanding of the "racial antecedents" of recruits and the overarching amateurism of the local draft boards. In the initial stages of the draft, it fell to the physician, "in whom the governor feels that he can repose implicit confidence," to carry out the physical examinations. With little to no understanding of the admittedly "imperfectly developed branch of knowledge that was physical anthropology," examining officers of the line were ill equipped to detect the apparent racial significance of all but the most apparent conditions, such as "deformities, skin eruptions, pallor, inebriety, flat feet, etc."[88] COA members noted that for recruits defined as "black," these defective conditions must be understood as endemic to, and not simply conditions of, an inferior racial constitution. By confusing physical deformities (such as flat feet) and infections with the far more insidious and entrenched forms of racial pathologies, draft board officials, as victims of cronyism and patronage, had failed to see the forest for the trees.

Debates regarding race selection and race betterment in regards to the draft intersected with the prevailing policies and practices of eugenics, the science of "better breeding" that informed much of the era's racial discourse. In the spring of 1918, in response to the revised procedures' state of apparent indifference to the "race question," Davenport received a civilian appointment in the War Department as an army anthropologist. Two months later, he was appointed Major in the Sanitary Corps. Finally, on July 23, 1918, a subsection of anthropology under Davenport's supervision was authorized within the Office of the Army Surgeon General.[89] Initially proposed as the section of "Anthropology and Eugenics," this subcommittee received federal funding and drew most of its staff from the COA. Hrdlicka predicted that anthropometric surveys of this kind would become "as fixed and important as the census, and the data they gather will serve as an index of progress, stagnation or deterioration of and within the nation, and thus be of vital importance to agencies for eugenics and legislation."[90] Drawing on Civil War anthropometry, COA officials remarked, "It seems as though we should, considering the progress of science, at least equal the achievements of Gould if not exceed them, and so the Committee has this purpose primarily in view."[91] Yet Davenport feared that there were limits to the utility of the Civil War-era data:

In view of the tremendous immigration of the past years, the physical changes of the racial constitution of our stock have been so great as to mask entirely any slight alteration that may have occurred in the physique of the stock of fifty years ago, through either improvement or deterioration of environmental or economic conditions.[92]

Contradictory statements such as these conflating hereditary and environmental factors—"racial constitutions" and "environmental or economic conditions"—only increased elite anxieties regarding the utility and explanatory force of eugenic terminology.

For much of the early twentieth century, eugenicists had dismissed African Americans as a factor in industrial civilization. Early eugenicists were primarily fixated upon class rather than race, positing degenerates and the poor as the greatest threat to the nation's biological well being. However, historians have noted that "the war accelerated the process of colored industrial advance" as southern black migrants began to replace whites throughout the industrial north.[93] Madison Grant—the patrician racialist and author of the best-selling lament for Nordic supremacy, *The Passing of the Great Race* (1915)—vigorously campaigned for a stronger eugenic element in the committee's work to better understand and combat the creeping "Negro factor" in national life. For Grant, an indifference to heredity was an indifference to history. Throughout their extensive correspondence, Davenport informed Grant that the inclusion of the term *eugenics* was to underscore the effects that the selection might have on the next generation in charting marriage and birth rates, and the "selective nature of the death rate in war and future matters of race and anthropology." However, Davenport maintained that he would withdraw the term if the Surgeon General should find himself "not educated up to it."[94] Eugenicists' fears that perhaps heredity was not in fact destiny, and that their goals would be misunderstood by the non-initiated, belied deeper anxieties regarding the mutability of racial physiologies during this highly volatile period of war and socioeconomic change.

From a practical point of view, eugenics—with its broad focus on the very production and ultimate prevention of certain degenerate types—was too unwieldy, too impractical for the purpose of racial labor management. Though keenly aware of the indispensability of "the Negro type" to the contemporary political economy, many observers began to doubt whether such a "type in its purest form" even existed. Prior to the outbreak of war in Europe, Grant warned Davenport about the dangers of airing these anxieties in public, given the current political and cultural climate. Responding to Davenport's musings on the prospective "whiteness" of mulattos and "Negroes as the first men," Grant remarked, "The inferior races are struggling so hard to assert their equality with the dominant races, that I think it a pity to give them any statements, which by misquotation could be twisted to their advantage. This is especially true of the Negroes, about whom so much sentimentalism has been wasted."[95] For Grant, race betterment was a zero sum proposition; white supremacy was contingent on blacks' continued debasement and subjugation. However, eugenic prescriptions for preserving white supremacy—whether state-enforced sterilization of the unfit, colored or otherwise—did not address the problem of the various degenerate types still at large in the present population. In contrast, anthropometry, with its ability to measure and quantify the respective utility of racial types, was seen as a powerful and effective tool for preserving white supremacy in both military and industrial labor hierarchies from the trenches of France to the shop floors of America.[96]

Establishment of a Subsection of Anthropology within the Army Surgeon General's Office (the term *eugenics* having since been dropped) was an attempt to craft a distinctly American form of anthropometry sensitive to the nation's racial and socioeconomic needs. A new section was authorized "to secure the highest quality of the measurements of recruits to assist the War Department in all questions about racial dimensions and differences."[97] The army hoped to answer these questions, cut down on rejection rates, and better prepare for the racial reconfiguration of the postwar workforce. During the summer of 1918, Davenport and his staff traveled to army camps across the nation to inspect methods and equipment used to carry out physical examinations. After a visit to Camp Grant in Illinois, they drew up revised plans—based on Hoffman's actuarial models, which traced links between disease and heredity—for use at the physical inspection fingerprinting at mobilization camps. However, due to the cessation of mobilization in September 1918 and the end of the war a mere two months later, these projects were never completed.[98]

Anthropometric studies of drafted men continued into the immediate postwar period. From the fall of 1918 until his discharge in January 1919, Davenport was relegated to a statistical analysis of the army's preexisting physical examination records. Davenport and his collaborator, Albert Love—both determined to guide future social policy along biological lines—rushed to publish their findings in *Army Anthropology: Physical Examination of the First Million Draft Recruits* (1919). Reviewing the data prior to publication, they found the "elucidation of defects to be highly subjective" and "shockingly indifferent to racial factors."[99] The data covered two successive drafts of Class I men from September 1917 to May 1918. Since the Class I men in the study were initially subjected to two different standards of physical examination, the results were predictably uneven. Nevertheless, the authors were able to make a few provisional conclusions regarding overall rejection rates. They found that recruiting standards declined over time; rejection rates fluctuated on a regional basis; and urban men were rejected at a slightly higher rate than those from rural regions. Sensory (eyesight) or physical (underweight) defects accounted for two-fifths (21% and 20%, respectively) of all rejections, followed by circulatory diseases (15.7%) and then tuberculosis and venereal diseases, which accounted for a combined 15%. Much to their chagrin, Davenport and Love conceded that their conclusions were bereft of any "real meaningful racial value" due to the committee's initial indifference to racial factors that consequently corrupted the final data.[100]

The failure of the Committee on Anthropology and the Surgeon General's Subsection on Anthropology led their respective members to explore new avenues well before their respective committees' official demise. Three initiatives in particular require attention. The genesis of the first began in March 1918, following a conference on eugenics at the Shoreham Hotel in New York at which Davenport suggested to Madison Grant the creation of a society that would analyze the race problem from the specific vantage point of physical anthropology.[101] This prospective organization would serve as a counter-weight to the American Anthropological Association, which had increasingly come under the influence of Boasian-trained anthropologists who privileged culturist interpretations of race over the biological models favored by the likes of Davenport and Grant. To reaffirm the society's commitment to hereditarianism, Grant suggested the name "Galton Society"—to honor eugenics founder Francis Galton—which met with Davenport's approval. Soon, Henry F. Osborn of the American Museum of National History—with whom Grant had co-founded the Bronx Zoo—was brought on board, and on April 16, 1918, a charter of the Galton Society was signed by its founding members. In conjunction with the Eugenics Record Office and the Museum of Natural History, the Galton Society would form part of a triumvirate of eugenic institutions based in and around New York City.[102]

Eugenicists' influence would grow in the 1920s reflected in the development of the Caucasian as a legal and cultural entity. New cultural forms of whiteness evinced in the era's "Nordic vogue" were disseminated to the nation through literary, cinematic, and intellectual channels emanating from the nation's cosmopolitan center on the island of Manhattan. For many old-guard native New Yorkers such as Grant, life in New York was akin to living in the belly of the beast, the focal point of the nation's descent into pluralist degeneracy.[103] One of the leading mediums in this new whiteness discourse was the new *American Journal of Physical Anthropology*, founded in early 1918 in New York and Washington, D.C by anthropologists and headed by Ales Hrdlicka. But Hrdlicka's COA peers, particularly Madison Grant, felt he was becoming indifferent to the wartime work of the committee. Grant believed that while the journal could be launched at any time, the window of opportunity provided by the draft into racial research was limited. Hrdlicka disagreed, and by the middle of 1918 he and Grant were no longer on speaking terms. Under the sole direction of Hrdlicka, the *American Journal of Physical Anthropology* blended Lamarckian and Mendelian theories of heredity. Hrdlicka proposed that environment "excited germ plasma" in various peoples, which exacerbated latent traits that individuals would invariably pass on to their offspring. Scholar Lee Baker argues that Hrdlicka and the journal were largely responsible "for making physical anthropology a well defined field within the discipline," a feat that culminated in the establishment of the American Association of Physical Anthropologists in 1929.[104]

The third and final alternative to COA refugees came through the creation of yet another NRC subcommittee, Race in Relation to Disease Civilian Records (RRDCR), founded in May 1918. The RRDCR was designed as a medium for linking military investigations of racial anthropometry to the public and private spheres. Frederick Hoffman was appointed head of the section and pledged to put Prudential Life Insurance's vast collection of vital racial statistics at the full disposal of the committee. Much like the COA and the Surgeon General's anthropological subsection, the RRDCR met with a mixture of bureaucratic indifference and intransigence from both the NRC and the federal government. Though the RRDCR had originally been conceived as single entity (incorporating both military and civilian records), objections from the Surgeon General's Office regarding civilian access to military records required the creation of two distinct military and civilian subsections with only loosely defined information-sharing rules—effectively eviscerating the committee's autonomy for much of its short existence.[105]

Measuring Race at the Eighth Wonder of the World

The RRDCR subcommittee was a key medium of wartime racial knowledge production, and approximated the focus of its committee predecessors to reconcile form with function. In correspondence with the Chairmen of the

NRC's Division of Medicine, Hoffman outlined three possible investigations: a study of Hawaiian race pathology with special reference "to Orientals"; "Racial Aspects of Autopsy Investigation" at John Hopkins Hospital; and "Race in Relation to Physical Condition and Fitness for Employment at Hog Island, Pa." The latter was intended to determine the physical variation of various racial groups that officials had so far failed to glean from wartime examinations.[106] Hoffman went over the records of Dr. J. J. Reilly, chief surgeon of the American International Shipbuilding Corporation at Hog Island, and found "the outlook for new and practically useful data as distinctly encouraging." He was especially impressed with Reilly's use of "photographic identification" in measuring a worker's stature, weight, and vital capacity. Despite Reilly's lack of formal anthropological training, Hoffman found him to be a man of "high scientific attainments with extended experience in the anthropometric measurements of men." Perhaps most importantly, Reilly was amenable to any of the committee's potential suggestions regarding policies and practices of measurement at Hog Island.[107]

Located southwest of Philadelphia along the Delaware River, the Hog Island shipyard was the largest in the world. Built in approximately nine months on reclaimed swampland, it was hailed by many as the "eighth wonder of the world." Foreign visitors to Hog Island described it as "awe-inspiring," "thrilling," and were thoroughly impressed with the "wonderful organizing efficiency of what they have seen." President Wilson stated, "Hog Island is simply wonderful." Lieutenant C. Wiezbicki of France succinctly expressed the conflation of war as work when, following a tour of the island, he remarked, "Hog Island is one of the two most important places in the world today. The other is the River Marne." Despite these accolades, the yard proved relatively ineffectual as an arsenal of democracy. The first ship produced at the site, the USS Quistconck was christened on August 5, 1918, but subsequent delays in its final fitting pushed its launch date to weeks beyond the armistice. In just over three years of operation (from August 1917 to January 1921), Hog Island turned out 122 ships equaling almost a million deadweight tons, yet failed to furnish a single ship for the war effort.[108]

Hog Island's importance resided in its seemingly limitless supply of human capital. The sheer numbers and diversity of men passing through the plant, comprising "dozens of different nationalities," made it an ideal source for gathering anthropometric data. Of an estimated 300,000 men, black laborers would make up approximately one-fifth of the island's workforce by war's end. COA officials noted that under the direction of the Industrial Relations Manager and the Consulting Sanitary Engineer, workplace hygiene and a general attention to working bodies "constituted one of the most important facets of life at Hog Island." Dr. Reilly commented, "We have not had an epidemic of any sort. Our record is better than any of the army cantonments and the spirit of the workmen is fine." The vitality of the workforce began with "the examination of all applicants for employment." Reilly's partner, Dr. Darlington remarked, "Efficiency means health. This country needs men too badly to waste them. Therefore, we are doing things here that are not usually done—we have a medical as well as a surgical hospital and vocational school on site."[109] A reporter for the *New York Times* noted that two large hospitals for employees were already up and running as well as a vast program of social welfare activities on behalf of the labor force—working bodies and the corporeal dimensions of labor efficiency retained pride of place at Hog Island.[110]

During the summer of 1918, NRC officials tentatively agreed to fund the RRDCR's proposed anthropometric research at Hog Island. The overriding question informing this work was the practical correlation between the physical variations of different racial types and their "physical efficiency, disease resistance, disease occurrence, physical, and pathological impairment."[111] From a military perspective, this investigation was intended to provide the War Department and the Office of the Surgeon General with "unquestionable evidence of physiological racial variations," which could help lessen the rejections caused by physical errors—thereby increasing the effectiveness of America's fighting forces. NRC and military officials hoped that anthropometric analysis undertaken at a key wartime labor site would provide insight into ways to preserve the industrial and racial equilibrium deemed so vital to a nation at war.[112]

Anthropometric methodology used at Hog Island was informed by a wider transatlantic anthropological discourse regarding the measurement and discipline of laboring bodies. Primary among these were models outlined by Charles Goring in his seminal study *The English Convict* (1913). Deploying an array of anthropometrical practices, Goring maintained that the criminal was a defective and not an atavistic physical type in the Lombrosian vein. For Goring, defective mental or physical traits were often concurrent, but not necessarily agents of degeneration: though not all defectives were criminals, most criminals exhibited defective traits.[113] Drawing on Goring's work, Hoffman believed that defective traits were generally inherited and could therefore be reliably traced through an analysis of the individual's "racial antecedents" visible in individual physiognomy. Though Hoffman and his peers were often unclear as to the existence of a priori "Negro type," they believed that certain measurable physiological traits—flat feet, concave chests, and flattened noses—collectively marked one as a Negro. Delineating racial difference in this manner effectively rendered race as an identity that was greater than the sum of its physiological parts.[114]

At Hog Island, maintaining workers' health began with the selection process. In early August 1918, Hoffman forwarded to Davenport the first draft of the subcommittee's record card for use at Hog Island. Hoffman hoped that "in the course of time, these cards could be found feasible for adoption on the part of many other industrial corporations or allied interests in connection with which employees or applicants for employment are physically examined." Therefore, "only such measurements and questions have been included that serves the immediate and practical purpose of determining the physical status, strength, aptitude and health of the person examined."[115] Examination cards were divided into four data groups: Personal Record, Birth Place and Race, Physical Examination, and Medical Examination. Aside from measuring an applicant's weight, height, and chest circumference, examiners were instructed to note any other "racial peculiarities" in lung capacity, feet, and hair color. Unless otherwise noted, recruits were to be measured in the nude. Finally, to increase the degree of accuracy and render any statistical findings compatible with those of European investigators, measurements were to be made in metric terms standardizing military efficiency across the Atlantic.[116]

Progressives were tied, if by nothing else, by an abiding faith in pragmatism through empirical inquiry as a tool of social reform. This instrumentalist view of ethics privileged systematic modes of theory and practice. Accordingly, Hoffman advised that different individuals be charged with securing different sets of anthropometric data in the interests of "systematic consistency." Examiners were urged to exercise great care in noting examinees' respective occupation and race. When filling in the occupational entry in Section A of the Personal Record, examiners were instructed as follows: "No general statements or assertions should be accepted. The mere term 'laborer' was deemed wholly inadequate." Applicants were to state the specific nature of their work—whether they were a laborer involved in sewer work, street-cleaning, or work at a steel plant—and their respective standing in these vocations. In Section B, "extra care was necessary for the ascertainment of an applicant's race," in an effort to link industrial aptitude with racial fitness. Sketching out these racial schedules, Hoffman asserted that "the large majority of applicants would be white, in the *ordinary* sense of the term."[117] This was a revealing designation given recent shifts in political economy such as black migration, cessation of European immigration, increasing labor standardization—all occasioning a shift away from heterogeneous ideologies of whiteness to a more homogenous Caucasian archetype.

Wartime demands privileged intra-racial over interracial models of racial difference. The proposed examination card also provided space for notation for Chinese, Japanese, and Filipinos. Hoffman suggested that it would be advisable to secure a list from the Bureau of Immigration of the "fifty principal races likely to be met with in this country."[118] Establishing a normative Caucasian archetype was ultimately an exercise in negation. Whiteness was defined not by what it was but by what it was not (that is, blackness). The physiological and social boundaries of this perpetually elusive whiteness required constant policing of suspected "deviant" and "abnormal" bodies through various legislative, economic, and social means. Moreover, this process—defined by distinct corporeal characters such as color and bodily shape—now became the primary markers of said deviance. When examining "Negro applicants at Hog Island, examiners would have to be guided entirely by personal appearance." The three "distinctions of 'black,' 'mulatto,' and 'near or almost white,' should not lead to confusion." For RRDCR officials, this was a tacit acknowledgment that black identity ultimately rested in its definitive non-whiteness. Anthropometrical investigations revealed that the "Negro type" was "colored" by a progressively declining absence of whiteness.[119]

The officials of the American International Shipbuilding Corporation were committed to a productivist ethos with little regard for caste or color. Black workers at Hog Island were initially given access to company health care and trained in highly skilled trades, such as welding and riveting. Thanks to the work of the RRDCR, however, opportunities for blacks were soon limited. During their time at Hog Island, Hoffman and Riley were able to evaluate approximately 450 workers through their new record cards. While the exact number of blacks in this sample is unknown, subsequent events reveal that they must have comprised a significant portion of the total. A mere six weeks after the drafting of the initial set of record cards, W. R. Wright, president of the Colored Men's Protective Association, brandished fifteen signed affidavits, claiming that black workers who had graduated from the Hog Island School as riveters had been assigned to jobs as general laborers due to their supposed "lack of aptitude."[120] International Shipbuilding officials countered that the promotion of skilled black laborers would only serve to antagonize white workers and disrupt the shipyards supposed racial harmony. Management cited Hoffman's anthropometric evaluations to buttress their claims of blacks' apparent lack of mechanical aptitude. The physiological imperatives of wartime production—the need to fit the right bodies to the right kinds of work—as opposed to callous prejudice, seemingly necessitated relegating blacks to the bottom of Hog Island's occupational hierarchy.[121]

Inadequately funded from the start and continually stonewalled by officials of both the American Shipbuilding Corporation and the NRC, the RRDCR was abruptly eliminated by the winter of 1918. Thus, Hoffman and his allies were unable to publish any of their preliminary findings on the Hog Island workforce. Following the armistice, appeals to the wartime "utility" of anthropometric work became moot. Like other initiatives of the wartime state, the U.S.'s brief wartime experience precluded any significant short-term results. The significance of the RRDCR was its ability to reconfigure socioeconomic ideologies in the long term and provide frameworks for future inquires into race and labor fitness. Through its association—albeit limited—with the American shipbuilding company, the RRDCR provided a template for strengthening links between the public and private sector "to establish a standard of physical efficiency for our soldiers and our adult male working population."[122] Anthropometry at Hog Island contributed to a broadening consensus that racial physiognomies could and should be measured and that a working knowledge of racial anthropology was vital to postwar national efficiency.

Wartime anthropometry was an applied science forged in the crucible of a nascent military-industrial complex. In the immediate postwar era, the COA and its various successor agencies prevailed upon leading labor organizations to note the racial consequences "of physical examinations in fitting applicants for employment."[123] Accordingly, Hoffman established a correspondence with representatives of the Western Pennsylvania Division of the National Safety Council and the National Industrial Conference Board regarding the broader implementation of these racial schedules. Mere weeks before the war's end, the committee achieved something of a coup, enlisting the support of American Federation of Labor (AFL) president Samuel Gompers who was currently serving as the chairman of the Committee on Labor at the Council of National Defense. Gompers was attuned to the discourse of industrial ethnography—developed by the likes of Ely, Commons, and Ross as a means to increase labor efficiency—and had long advocated ethnographic inspection of workers. Industrial ethnography was especially appealing in that it provided a scientific rationale for the AFL's advocacy for skilled labor, an aristocracy of labor whose elite status could seemingly be reified through an analysis of working bodies. But Gompers also recognized that a complete ethnographic examination of every prospective laborer, however desirable for medical, public health, or race betterment purposes, was beyond the financial and institutional means of most companies.[124]

Notwithstanding the lack of financial and institutional resources, many managerial elites opposed industrial anthropometry on principle. Following the Civil War, any large scale public and private investigations into racial physiologies were invariably bedeviled by the fear that a lack of normative racial taxonomies would allow various racial types to pass undetected into positions or occupations for which they were not "racially equipped." The practice of "passing"—whether in the social or industrial spheres—revealed longstanding anxieties on the part of whites and blacks regarding the mutability of the color line. For Gompers and his allies in labor and industry, the anthropometric record card was seen as a way to efficiently rationalize working bodies in explicitly black and white terms eliminating any potential ambiguities. These practices would do away with time-consuming medical examinations. The record card would far more effectively chart bodily development, nutritional conditions, and occupational aptitudes than the medical literature currently used by industrial agencies—which was generally indifferent to racial dynamics. By the spring of 1919, the COA was inundated by inquires for racial record cards from corporations such as Standard Oil, Sears Roebuck, and their former partners at Hog Island, the American Shipbuilding Corporation—all seeking to reconcile racial form with function.[125]

Race and Postwar Sartorial Ergonomics

Postwar efforts to link certain bodies to certain work coalesced along sartorial lines via the ready-made clothing industry. In the winter of 1918, Harvard anthropologist Dr. E. F. Hooton wrote to Secretary of War Newton Baker citing the "tremendous problem afforded by large numbers of the Negro race in this country." Hooton warned that "to neglect the investigative opportunity presented by demobilization would hamper the scientific study of race betterment for many years to come." Hooton's plans received a boost during the following summer when the Secretary of War directed the Office of the Army's Surgeon General (OASG) to have "measurements of 100,000 men made for the purposes of uniform patterns."[126] With the final authorization of Col. A. J. Dougherty of the equipment branch of the OASG, a cadre of expert anthropologists under the guidance of Dr. Charles Davenport was enlisted to perform this work throughout the various national camps where demobilization was already under way. Given public demands for a quick return to "normalcy," anthropologists were cautioned to finish their work quickly "so as not interfere in any way with the process of demobilization" in keeping with wider public demands for a quick return to an imagined prewar stability.[127]

Sartorial anthropometry is an often-overlooked form of racial labor management of the immediate postwar era. Fitting men for military uniforms or industrial clothing—"accruements vital to their respective industrial productivity"—allowed Davenport and his peers to appropriate the impetus of wartime imperatives for a postwar civilian industrial context and to reconstitute anthropometry as an applied science essential to national efficiency and race betterment. COA anthropologists were tasked with collecting data to be used "in the construction of manikins of various sizes with the aim of affording better fitting uniforms for the army." However, the COA's primary intent was to ascertain the diversity of racial types, since "the tariffs of sizes to be supplied to any distribution zone for a draft army depended on the racial constitution of the population living in that zone."[128] These prized measurements were obtained under the pretext of improving military efficiency, so as to not "unduly alarm any uncooperative radicals who might object to being measured for eugenic purposes."[129] Regarding the formation of the two all-black combat divisions (the 92nd and the 93rd), officials claimed that "surely the question of whether a given person had Negro blood must often have arisen and mistakes have been made." These suspected incidents of "passing" were believed to exert potentially disastrous effects on the biological composition and integrity of a unit. Through its sartorial manufacture of bodily types, the ready-made clothing industry was the ideal medium to expedite the process of linking race and labor efficiency through the body.[130]

The cadre of postwar anthropologists comprised a large part of those individuals involved in wartime testing. One month after the Army Surgeon General placed its order for uniforms, Davenport tapped Ales Hrdlicka, now the curator of the division of anthropology at the National Museum in Washington, D.C., to train anthropologists for work in the demobilization camps. Under the direction of the camp surgeon, anthropologists were required to measure up to approximately ninety men an hour per eight-hour day. Given the limited allotted assignments for each anthropologist and the basic cost of maintaining men in the service, "haste was essential" to ensure that this unprecedented opportunity was not lost. Examinees were to be measured in the nude using apparatuses such as calipers, measuring tapes, and metal scales. There were twenty primary metric measurements ranging from standards such as "standing and sitting height," "weight," and "circumference of chest" to more obscure notations of "height of pubis" and "height of sternal notch." Final measurements were shipped to the Medical Record Section of the Surgeon General's Office in Washington, D.C. for transfer to Hollerith punch cards using a prearranged code for bodily measurements.[131]

Wartime and postwar anthropometry inextricably linked race to color. Visual subjectivities trumped all other forms of sensory analysis in defining who or what a Negro was. Officials were advised to use their "own discretion to judge the fraction of Negro blood by estimate of skin color." Accordingly a "mulatto" was identified by their "clear brown" or "cafe au lait" complexion. If the skin color was darker than clear brown, one was to mark them as three-quarter black. If the skin color was light brownish, yellow, or lighter ("yet clearly of African Descent," but presumably marked by other physiological markers such as hair or eye color), the examinee was defined as one-quarter

black. Conversely, other veterans of color were given unprecedented opportunities to racially self-identify. In the case of a "probable" or "suspected" individual of Indian, Chinese, or Japanese descent, officials were simply advised
to ask: "Of what race?"[132] The measurement card also contained blanks for
recording the "native language" and "religion" of the veterans' parents along
with the "nationality" of his maternal and paternal grandparents. If this line of
questioning proved insufficient, examiners were provided space to note any
"other noteworthy racial traits" that could provide clues to the individual's
"precise" racial identity. Whereas other colored peoples were given a chance
to self identify, those deemed "Negro" were afforded no such opportunity. For
wartime officials, blackness was a fixed identity firmly rooted and identifiable
in the body.[133]

COA anthropologists found that their sample size of 100,000 men—of
whom only 6,445, or a mere 0.6% of the total, were deemed "colored"—was
too limited to glean any meaningful conclusions about recruits' labor fitness.
Consequently, the demobilization material was combined with prior wartime
evaluations of the first one million recruits and published as *Army Anthropology* (1921) by the Army Medical Department. Authored by Charles Davenport
and Albert Love, *Army Anthropology* detailed the sizes and proportions of the
American male population from twenty-one to thirty years of age with "reference to health and development, to geographical distribution and environment, and to race and color." With a larger sample size, the authors were able
to make some definitive conclusions regarding racial physiology. The general
comparative picture of white versus black troops revealed the latter as having longer appendages, shorter trunks, broader shoulders, narrower pelvises,
and greater girth of neck, thigh, and calf than their white peers. While blacks
seemed more powerfully developed from the pelvis down, whites were more
powerfully developed in the chest—confirming longstanding beliefs in white's
seemingly superior respiratory health as a key indicator of labor fitness.[134]

Postwar anthropometry was a mixture of heredity and environmental
models of racial development. Davenport and Love mused as to what role,
if any, environment or acquired characteristics played in the development of
racial types—and could a shift in function potentially alter form? The primary aim of *Army Anthropology* was to determine the effect of the military
experience on racial physiologies.[135] The authors concluded that it was "important to know the physical characteristics and proportions of accepted and
demobilized men, as physical proportions are intimately related to mental development, diseases, and nutritional requirements." Therefore, all measured
men were required to have a minimum four months of military service to
better gauge the role—if any—of military service in regard to recruits' physi-

ologies. When it came to blacks, these evaluations could aid in charting "the changes which have occurred in the Negro due to various conditions and the persistent qualities which occurred in the distribution of the race during its recent northward migration." Officials noted, "In three primary southern demobilization zones, respectively 35%, 30%, and 25% of the men measured were to be of African descent." Measurements were then matched to the available census data to effectively provide officials with a corporeal map of black proletarianization.[136]

Postwar data confirmed the physiological effects of military service on recruits of all races. From enlistment to demobilization, the mean stature of all enlisted men, both white and black, had increased from 67.49 to 67.72 inches. Blacks stood at 67.70 inches, compared to 67.71 inches for whites, while the average weight of all military personal increased by approximately three pounds, to 144.50 pounds. The reasons for these changes were numerous. Young men of all races had lied about their age at enlistment, and thus underwent slight growth spurts while in the service. And height and weight requirements had been progressively lowered in later drafts to redress manpower shortages. Yet, despite sharing a near identical height with whites, blacks emerged from the war five pounds heavier on average than their white counterparts. Davenport and Love also noted that "in the southern sections, those containing many colored men show relatively less obesity than those containing a small proposition of them."[137] Finally, while the chest girth of black troops was somewhat less than that of whites, the chest circumference of recruits had increased about one inch over the last twenty-four months of military service, from 33.22 to 34.94 inches. The authors concluded that, thanks "to the fine physical conditions of army life," the average recruit emerged from the war a changed man: a slightly taller, heavier, and fitter version of his former self. For many recruits, especially those from overcrowded urban areas, these fitter physiques were due to little more than increased access to outdoor activity, exercise, and better nutrition.[138]

Postwar anthropometry yielded key (yet often contradictory) insights into the shifting nature of racial types. The editors of the *Journal of the American Medical Association* cited *Army Anthropology*—along with an additional 549,099 men who were rejected by local boards as totally and permanently unfit for military service—to argue that "the uninfected Negro was a constitutionally better physiological machine" than his white counterpart.[139] This assessment ran counter to prevailing theories of blacks' mental inferiority gleaned from the infamous Army IQ tests. The U.S. Army and the American Psychological Association (APA) administered the IQ intelligence tests under the direction of Robert Yerkes. Historians have demonstrated how IQ tests linked a suspect regime of assessment to the national interest and featured "wildly biased questions in terms of race, nationality, and class." Examples included queries that would have been baffling to the majority of the population, such as, "Why is tennis good exercise?" and "Where is Cornell University?" Testers gave recruits a letter grade and while "new immigrants" such as Italians and Poles consistently registered in the "D" grade range, black Americans represented a negative baseline of "F" against which all others were measured.[140] An astronomical 89% of African American recruits, compared to 47% of whites, were classified as "morons," defined as possessing a mental age of seven to ten years. Officials ignored social disparities between Southern and Northern blacks test scores—with the latter scoring consistently higher—as well as the disparity of educational rates between whites and blacks. While 25% of all draftees were classified as illiterate, the median number of years spent in education ranged from 6.9% for "native" whites, 4.7% for immigrants, to only 2.6% for Southern blacks. Reflecting on these tests in the immediate postwar era, one veteran quipped that many officials "take these soldiers who whipped the Germans to be mental and physical cripples."[141] Notwithstanding their social biases and shaky methodology, the tests further convinced eugenically minded Americans not only that mental deficiency was genetically determined but also that physical *and* mental function were linked.[142]

Wartime anthropometry told a different story from a physiological perspective. Physiologically, blacks and white shared a similar mortality rate for the majority of illnesses, though the black solider "possessed more stable nerves, has better balance, and metabolizes better" than his white counterpart. The skin of black soldiers—on the bodily surface, and in enfolds of the mouth and nasal passages—was found to be much more resistant to microorganisms than that of whites. Officials found that white skin actually seemed to "be a relatively degenerate skin in this respect."[143] Blacks suffered from few of the diseases of "over-civilization," such as neurasthenia and various other psychopathic disorders, due to their allegedly innate primitivism. If "civilization was paid for by nervousness," as George Beard, author of the seminal text on neurasthenia, *American Nervousness* asserted, then those deemed the antithesis of civilization, such as blacks, were absented from this transaction. In this vein, blacks seemed to lack the requisite nerves to be frayed, or minds to lose, in the first place.[144] Despite high levels of poverty, blacks were also less likely than whites to suffer from nutritional disorders or alcoholism—issues that affected whites at nearly double the rate of blacks. However, blacks were found to be more susceptible to maladies such as venereal diseases, tuberculosis, and smallpox. The apparent physiological superiority of the black soldier was all the more impressive given a selective service that often enlisted unfit black recruits without causing any apparent diminution in the health of the seemingly superior Negro type.[145]

Persistent attempts of the COA to compile working taxonomies of racial labor types were constantly hindered by a sustained lack of institutional support, bureaucratic incompetence, and committee infighting. Yet those studies yielded some perplexing and unsettling results. Perhaps most troubling for white social scientists, the allegedly primitive and ossified figure of the Negro had displayed a surprising capacity to work his way to a state of civilized fitness. Perhaps the experience of Henry Berry—transformed into a fitter and healthier man through military labor—had proven the rule rather than the exception. White elites found the wartime emergence of the Negro as a "superior physiological machine" troubling on many levels, not only because it refuted notions of black inferiority, but also because it challenged the stability and utility of black and white racial identities. More worrisome than blacks' transformation into a supposedly superior type was their ability to transform themselves in the first place. In the face of blacks' apparent physiological superiority, social scientists were forced to rethink their fundamental understandings of immutable racial typologies. Perhaps racial form was changeable. Regarding army anthropometry, Davenport maintained: "We started this draft in ignorance; all our errors were due to the fact that we had a heterogeneous mass of data instead of a homogenous distribution of types."[146] Flawed methodology inevitably produced flawed results. However, he remained optimistic about how the tools of wartime anthropometry could be used to maintain, or reestablish, racial labor hierarchies in the postwar era. In a lecture to the American Association for the Advancement of Science, entitled "The Measurement of Men," Davenport noted:

> Man is more than a physical body. A live person has a dynamic output, and it is one of the signs of progress in the scientific study of men that within the last fifteen years so great an extension of the measurement of human functioning has been undertaken . . . It led to the remarkable intelligence testing and physical measurements in the late war and the attempt to express quantitatively the social and physical qualities amongst army recruits.

But he concluded with a warning: "Although type is a natural refuge of the mind, the danger is that in our enthusiasm for selected types the great mass of people who do not fall into any type are overlooked."[147] In the end, the soldiering body across the color line proved remarkably resistant to taxonomic standardization.

The carnage of war and increased standardization and segmentation of wartime labor processes had violated the productive integrity of the working body. Reconstituting racial typologies as a dynamic output of physical and mental forces allowed social scientists to rationalize the apparent physiological superiority of the Negro by calling into question the very parameters and contours of the Negro type. Laboring bodies—and by extension, labor power—was reduced to little more than the sum of its parts. Though some of these parts were seen as interchangeable—witness the growth of the postwar prosthetics industry—others were thought to be firmly rooted in heredity. The Army IQ tests that labeled blacks as "moronic" justified white presumptions that African Americans' ostensible physiological superiority was invariably compromised by their inherently feeble intellect. Though the shape of the Negro may change, his essentially debased intellect was fixed. Whereas whites possessed the genetic capital to overcome their particular mental and physiological shortcomings, the Negro was undone by his inability to reconcile his own. The mind-body tension that undermined his "dynamic productivity" left his savage—yet vital—body a hostage to his primitive mind.[148]

Biologically informed racial typologies persisted as key organizing principles of labor economies notwithstanding the creeping culturist mindset of the postwar era. Anthropometry continued to be described as "the best quantitative expression of the form of the body" through much of the postwar period. For Davenport, "wartime anthropometry was a handmaiden to genetic studies directed toward problems of inheritance of the human form in both the military and industrial sphere."[149] Concurrent with postwar assessments of wartime anthropometry was the practice of rehabilitation and the struggle to determine which individuals or groups required or warranted reconstruction. Both anthropometry, and later, rehabilitation sought to determine the value of laboring bodies, albeit in differing stages of bodily integrity. Theories of black physiological superiority gleaned from anthropometric inquiry did not go uncontested during the various stages of rehabilitation. This continued as officials struggled to determine whether they could or even should mend black bodies that they already assumed to be broken. The tension that informed social scientific efforts to simultaneously define and rehabilitate a supposed "Negro type" within the confines of legalized and de facto segregation marked a turning point in progressive imaginings of the black worker and the very nature of work itself.[150]

4

Salvaging the Negro

Vocational Rehabilitation and African American Veterans, 1917–1924

"It is extremely difficult for the opposite race to see the colored
soldier in a fair and impartial light. The fact is that we are invariably received and treated as a colored man and not as a disabled
soldier."
—Private James Sanford to Dr. J. R. Crossland, Veterans Bureau
Negro Advisor, 1921

In August 1922, Buster Sunter, an African American veteran of World War
I, informed the Veterans Bureau of alleged mistreatment at his local veterans training center. Previously diagnosed with tuberculosis, Sunter wrote,
"I want to let you know that I have not been treated right here, when I take
this training I was supposed to have four years here but they have cut my
time down to two years . . . they bully me and have me work like I ain't
sick so I want you to look into the matter for me for I am not able to work
like that." The experience of Private Sunter—of officials refusing to treat, or
even acknowledge his condition—reveals how ideas of race and disability
shaped the policies and practices of rehabilitation in early twentieth-century
America. Postwar models of disability were both medical discourses and
social constructs, developed by physicians, legislators, administrators, and
veterans within the broader networks of labor, gender, citizenship, and race.[1]
The cultural politics of remaking men for work in the "great industrial army"
fused the health of ex-servicemen with that of the republican body politic in
explicitly racial terms.[2]

Modern war brutalized bodies in shockingly new and gruesome ways. The
First World War's human cost was staggering: 8.5 million dead, 20 million
wounded, and more than 8 million permanently disabled. Despite being a
latecomer to the conflict, the United States lost 116,708 men, with an additional 210,000 classified as wounded or disabled. The disabled came in many
forms: mentally ill (victims of "shell shock"), tubercular, syphilitic, blind,
deaf, and amputees.[3] The mechanized murder of the trenches irrevocably

shattered Victorian ideals of heroism and war as a rejuvenating force for white Western masculinity. For Lothrop Stoddard, author of the bestselling work, *The Rising Tide of Color Against White World-Supremacy* (1920), the war was nothing less than "a catastrophe" because its "racial losses were certainly as grave as the material loses."[4] For Stoddard and many of his racialist peers—those who saw race as the prime engine of historical change—the carnage of war and its ensuing economic and physical "waste" did not occur in an ideological vacuum. Historian David Gerber has noted that all soldiering bodies "were endowed with signs and declarations of age, generation, class, ethnicity and race" and within these frameworks, bodies lived, died, and were broken.[5]

Rehabilitation was a deeply political process that challenged prevailing ideologies and identities of race, gender, and work. But as Gerber reminds us, if "war is the extension of politics by other means, then only by making victims can war achieve its political aims."[6] The issue of rehabilitation ultimately rested on an assessment of the state's obligation to its veterans. Either the state would secure the employment of tens of thousands of injured men, or it would abandon them to an unpredictable marketplace, and the pitying attitudes of their fellow citizens. Following World War I, science displaced sentimentalism as the core principle of veterans' care. Throughout the West, the question of rehabilitation was "a theoretical, an economic, and finally a moral one, but its consequences for veterans were eminently practical." Veterans groups argued that, "if the state had the power to draft men, it also had the responsibility to prevent the war from ruining the lives of those it conscripted." Loath to encourage an emasculating culture of dependency, officials pledged that, "the government is resolved to do its best to restore him (veterans) to health, strength, and self-supporting activity." Intellectuals such as John Dewey worried that "despite the struggle for the elimination of Prussianization, we are at the same time secretly admiring and envying it."[7] The reformist impulses of rehabilitation through the pursuit of physical perfectionism were ultimately constrained by America's unique racial dynamics and historical distrust of an activist federal state.

Though the war did not create the impetus for state surveillance, it intensified and encouraged the proliferation of new regulatory institutions and practices such as the draft and vocational rehabilitation. One such institution, the Federal Board of Vocational Education (FBVE) was charged with rehabilitating the citizen-solider into the citizen-worker. FBVE officials developed catalogues of racial labor taxonomies to delineate which bodies could do which kinds of work. Rehabilitationists, fearful of rising labor radicalism, racial unrest, and increasing industrialization, sought to mend the social fabric one individual at a time. Yet, through the stages of diagnosis, job training, and hospitalization, officials struggled to determine whether they could, or even should, mend broken black bodies often understood as defective by definition.[8] FBVE policy devalued and institutionalized disabled African American veterans and dismissed their claims to rehabilitation as spurious attempts "to unjustly profit from their natural inferiority." Black ex-servicemen resisted these processes and contested their right to rehabilitation as soldiers, citizens, workers, and men.[9]

Rehabilitation sought to reconcile physiological form with labor function to distinguish the "deserving" from the "undeserving" disabled. Throughout this process, FBVE officials drew on prevailing theories of industrial evolution to connect racial development with labor fitness.[10] Analysis of the FBVE's policies towards African American veterans provides key insights into the production of interwar racial labor hierarchies, the rise of racial expertise and the drive to shape social policy along biological lines. This trend culminated in 1924 with the passage of the Johnson-Reed Immigration Restriction Act and Virginia's Racial Integrity Act—the latter banned inter-racial marriage and authorized involuntary sterilization of the mentally and physically "unfit." Historian Matt Price describes rehabilitation as an utopian exercise in which "many broken threads, representing physical, mental, and social factors, must be unraveled and rewoven to make a consistent pattern" of national health.[11] Rehabilitation was fundamentally an exercise in individual and social perfectionism that was beholden to contemporary ideologies of gender, work, and race.

Rehabilitation also provided a conduit between national productivity and the preservation of racial integrity. The drive to remake broken black bodies was driven by a segregationist impulse to put the Negro in his proper occupational and social space. While the war shattered popular notions of progress, rehabilitationists sought to reconstitute Western civilization from the conflict's human detritus. The soldiering body, which had once been viewed as a source of social contagion and vice, was reconfigured by the FBVE following the armistice—the crippled soldier's body was now a source of national and racial regeneration. Reformers such as Elizabeth Upham argued that "the wide prevalence of defects found through the physical examinations of the draft justifies a careful consideration of physical and mental racial health when developing programs of reconstruction."[12] These medical models of disability were informed by the assumption that "pathological physiological conditions [were] the primary obstacle to disabled people's social integration."[13] Elites conceded that while rehabilitation could remake the near-white immigrants of eastern and southern Europe into dutiful, efficient Americans, non-white peoples such as Asian, Native, and African Americans required social and oc-cupational segregation to protect them from themselves and society at large. White elites were increasingly disturbed by the fact that the "Negro was fast becoming a factor in the nation's industrial landscape."[14] Between America's entry into the war and the stock market crash of 1929, African Americans left the South at an average rate of five hundred per day, or more than fifteen thousand per month: transforming an almost exclusively rural populace into an increasingly urban one.[15]

Following America's entry into the war in April 1917, prevailing anxieties regarding blacks' lack of martial fitness denied most a combat role. Army officials generally believed that "the poorer class of backward Negro has not the mental or physical stamina or moral sturdiness to put him in the line against opposing German troops of high average education and training."[16] Only thirty-eight thousand blacks served in overseas combat units, con-stituting only 3% of the army's combat forces.[17] The vast majority of black recruits were relegated to segregated service battalions laboring at the in-glorious work of war as stevedores, cooks, and menial laborers. White of-ficers assigned to these battalions often treated their charges as little more than beasts of burden and constantly endeavored to "work the hell out of the niggers."[18] After the war, tens of thousands of black troops were charged with the gruesome task of exhuming the American dead for re-internment in France or the United States. Blacks' physical absence from the battlefield led to their figurative absence from postwar debates over the rights and re-sponsibilities of rehabilitation.[19]

Emboldened by wartime rhetoric espousing democracy and self-determination and exposed to rising anti-colonial sentiment among African colonial troops in Europe, many black troops returned home determined to pursue full equality. W. E. B. DuBois gave voice to this ethos in eloquent terms, perhaps in part to absolve himself of his earlier pro-war leanings: "We return, We return from fighting, We return fighting; Make way for democ-racy! We saved it in France, and by the Great Jehovah, we will save it in the United States of America, or know the reason why."[20] Whites—especially those in the South—reacted to blacks' newfound stridency with a mixture of bewilderment, indifference, and murderous violence. Reported lynchings rose throughout the nation—from sixty-four total in 1918 to eighty-three (76 of whom were black) in the bloody summer of 1919. Black veterans were popular targets for these most extreme acts of racial vigilantism; at least two dozen fell victim to the noose—often while still in uniform—during this pe-riod.[21] Monroe Work, editor of the *Negro Year Book*, summarized the chal-lenges facing the South, and indeed the nation, in the immediate postwar years, by "first the handling of demobilization in such a way to prevent racial friction or conflict; second the maintenance of those harmonious relations that have already established." Despite Work's dubious claim to prior "harmonious" race relations, he did echo a widespread belief among progressives that racial concerns were a key component of postwar Reconstruction. Work's prescriptions for racial reconstruction, however, belied a profound anxiety shared by many of his peers regarding blacks' transition from citizen-soldiers to citizen-workers. As he wrote in the pages of *The Survey*, "All the men seem glad to be home again, and on the surface at least to accept their social inferiority as a matter of course. What goes on in their minds beneath that surface cheerfulness and docility no one seems to know exactly?"[22] Rehabilitation sought to make the increasingly unknowable Negro of the postwar progressive imagination knowable through the discipline and reconstruction of the black soldiering body.[23]

The Roots of American Vocational Rehabilitation

Shortly after the outbreak of hostilities in August 1914, the modern science of rehabilitation came to the fore through the efforts of a coalition of social, medical, and scientific reformers. Historians have noted that "the military requirements of modern warfare and industry provided governments with a powerful incentive to intervene in new areas of the economy including the construction of men's bodies."[24] Advances in medical and surgical care in fields such as orthopedics meant that more men could be "salvaged" than in previous conflicts. Throughout the Western world, "aggressive normalization" through physical restoration and vocational education was conceived as a balm to the dysgenic effects of modern warfare.[25] Prewar and postwar models of rehabilitation shared a distinctly naturalistic character evincing the era's overarching attempts to view social dynamics through biological lenses. As early as 1909, Herbert Croly, in *The Promise of American Life*, had argued for a national system of vocational education, "so that the laborer be placed, just as timber, stone and iron, in the places for which their natures fit them."[26] Just less than five years after the armistice, the editors of the FBVE's official organ, *The Vocational Summary*, remarked: "Conservation of our natural resources has been one of the most important developments of the twentieth century. We have reclaimed our arid lands; we have plowed our burned forests into fertile fields; we have taken our discarded metals from the scrap heap and remolded them to other uses. By a natural evolution, crystallized by the casualties of war we have come to the problem of salvaging our men."[27] Lieutenant Henry Mock of the Army Sanitary Corps cited "scientific human conservation" as one of the "greatest byproducts of the war."[28] By war's end, the drive to sustain the best in nature coincided with a desire to conserve the best of humanity.

Rehabilitation was a transnational project, uniting reformers and elites from London, Paris, Toronto, and New York. Americans moved quickly to integrate themselves into these transnational networks even prior to their official entry into the war. Officials lamented that "no pioneer road was left for us to follow with respect to the physical reconstruction and vocational rehabilitation of our wounded . . . it was left for us to merely select a plan and to modify it to our own needs." Consequently, American rehabilitationists drew heavily on British, Canadian, and even German models of reconstruction. Due to Canada's proximity and prior experience with close to three years of war, the cities of Toronto, Montreal, and Halifax became key stops on Americans grand tour of foreign institutions of physical reconstruction. Americans were especially keen to expand on Canadians holistic forms of rehabilitation by blending physical with vocational reconstruction to facilitate the "training of the disabled man to again be a productive agent in spite of his handicap." Major John Todd of the Canadian Pension Board envisioned rehabilitation as a communal practice "a matter of such wide extent that it can leave no phase of social organization untouched."[29] This was part of a broader attempt to substitute independence for dependence and move veterans' care from the realm of the sentimental to that of the scientific—remaking both the individual and the body politic writ large. Douglas McMurtrie, along with W. M. Russie, co-founded the Institute for Crippled and Disabled Men in New York City in 1917 and framed rehabilitation, the nation's "duty to the war cripple," as an act of wholesale social regeneration. He urged surgeons in the various frontline base hospitals "to free themselves from their tendency to treat the wounds and forget the function; to make a well man but not a working one; to take the anatomical rather than the physiological point of view." McMurtie believed that the "only compensation of real value for physical disability is rehabilitation for self support. Make a man capable again of earning his living and the chief burden of his handicap drops away." FBVE member Dr. W. S. Bainbridge argued that "the rehabilitation of the soldier is a redemptive act for the nation—it demonstrates a nation's moral fiber and authorizes that nation's economic success": morality and corporalities reimagined as a function of national efficiency.[30]

Following the U.S.'s entry in the vocational rehabilitation became law with the passage of the Smith-Hughes Act in February 1917. The act called for federal support to "train people who have entered upon, or are preparing to enter upon, the work of the farm."[31] Administration of the act at the state level was facilitated through a Federal Board of Vocational Education (FBVE). The FBVE expanded on the agricultural focus of the Smith-Hughes Act by empowering the states to promote vocational education in agriculture, home economics, and industry. To secure appropriations for teaching and institutional support, states were required to accept the provisions of Smith-Hughes through their legislatures or their governors. Moreover, states had to designate or create a board of at least three members with the necessary power to cooperate with the FBVE at the local level.[32]

Developing provisions for the rehabilitation of disabled soldiers and sailors was originally only one of the FBVE's duties, but soon became its primary function. Beth Linker notes that "unlike Europe and Canada, where rehabilitation was largely a voluntary component of a disabled veterans benefits package, the U.S. compelled disabled servicemen to undergo long term medical treatment."[33] In January 1918, a bill was presented to Congress recommending rehabilitation work be administrated by a commission of five persons representing the Office of the Surgeon General, War Department, Department of Labor, Bureau of War Risk Insurance (BWRI) and the FBVE. On June 27, 1918, the Vocational Rehabilitation Act was passed with a federal approbation of approximately $2 million. The act provided for the vocational training of "persons disabled in the military or naval forces of the United States, and for their assistance in obtaining gainful employment" following their discharge from the service. Although the FBVE initially focused on restoring veterans' disabled bodies, it soon extended to the conservation of "national energies" by providing services to civilian victims of industrial accidents—furthering the model of war as work.[34]

Postwar rehabilitation policy was a direct response to previous veterans' programs characterized as inefficient, wasteful, and corrupt. At the turn of the century, over a million Americans were receiving pensions totaling $150 million a year, approximately 38% of the entire federal budget. By 1917, the federal government had spent $5 billion on military pensions since 1776, with the majority going to Union veterans of the Civil War. Progressive reformers wedded to the cult of efficiency, believed that the aging veterans of the Grand Army of the Republic were a drain on federal resources and the national work ethic. To prevent the swelling of this "dependent army of cripples," the Vocational Rehabilitation Act stated that only "severely disabled" soldiers who qualified for compensation under the War Risk Insurance Act were entitled to retraining.[35] One FBVE official bluntly defined the bureau's policy as "one of conservation," adding that "in the treatment and placing of disabled men back into industry, there is no room for the spectacular."[36] Dr. W. S. Bainbridge of the FBVE employed slightly loftier rhetoric: "The rehabilitation of the soldier is a heroic and redemptive act for the nation—it demonstrates a nation's moral fiber and authorizes that nation's economic success."[37]

The FBVE was committed to expunging the culture of victimization from veterans' policy. However, the relatively low educational and physical status of the average Doughboy—manifested through wartime intelligence and anthropometric testing—lowered the expectations of many FBVE officials. Consequently, FBVE models of rehabilitation vacillated between the pragmatic and the utopian. Many of the bureau's personnel saw vocational rehabilitation as a means to transform the "lucky handicap" into a more productive version of his former self. The editors of *The Vocational Summary* noted, "Practically every man, no matter how handicapped he may be, can come back. In fact a handicap puts more fight into a man, makes him strive harder than ever before, and results quite often in his making good to a greater extent than if he had never been disabled."[38] Willpower was essential to escaping the dreaded state of deformed dependence. Officials triumphantly cited the case of a double amputee who remarked, "'Watch me! I am going to make good with both feet.' And he has. This is the spirit! Determination and grit—stick-to-itiveness—are the qualities which every disabled man must have or must acquire to crawl out or jump out of that hated class—The Disabled."[39] Only by embracing "a sober program of reconstruction" could veterans achieve a manly self-sufficiency. Most importantly, veterans had to be made aware of the "naked grim reality" facing them after all of the "well meaning sentiment and verbiage about heroism and gratitude and never forgetting" faded away.[40] The FBVE initially placed veterans in educational institutions as opposed to on-the-job training to insulate them from the depressed wages and rapid shifts in industrial labor processes of the immediate postwar years. They feared such reality could potentially undermine the ethos of self-sufficiency that bureau officials were attempting to inculcate in their charges.[41]

The policies and practices of rehabilitation were also deeply informed by the era's gendered politics. In the spring of 1918, the returning waves of crippled men and the rehabilitation they required gave reformers a chance to push women and blacks out of their "temporary" industrial jobs and back into the "ennobling protection of the home and the field." Efforts to restore white male dominance in the labor market were based on the belief that because "the life of a nation springs from its motherhood," women required protection from the debilitating environment of the shop floor. Despite women's increasing role in wartime industries—an increase of approximately 20% increase from 1910 to 1918—many employers were hesitant to retain them in the postwar era.[42] Dr. Francis Patterson, chief of the Division of Industrial Hygiene for the Pennsylvania Department of Labor, claimed, "While it is undoubtedly true that women can and are replacing men in some positions, by reasons of their sex there are limitations upon the work they can do. It needs no words of mine to emphasize the importance of the conservation of the health of those who are to be the mothers of our future race." George Lipsitz notes that "patriotism has often been constructed in the United States as a matter of gendered and racialized obligation to paternal protection of the white family."[43] In the immediate postwar years, these obligations were reconfigured as a function of national efficiency.

Mass migration of African Americans to the industrial north along with women's wartime entry into industry necessitated a reimagining of their respective fitness for modern industrial work. White elites in both the public and private sphere were generally forced to concede that African Americans' wartime industrial labor had proven to be "quite as good as the foreign labor" that it had replaced.[44] Yet in order to restore traditional gender and racial hierarchies in the industrial sector, FBVE officials could no longer simply cite the negative effects these environments had *upon* the "female and negro character." Instead, they had to rely on locating female and black workers' deficiencies *within* their respective physiologies. The weak postwar economy, increasing labor standardization, and the socioeconomic demands of white patriarchy required merging form with function to delineate the limits of gender and racial labor fitness.

Rehabilitationists focused as much attention on conserving the physical output of bodies—the natural kinetic power crucial to economic success—as they did on efforts to restore battered male psyches. The wartime "Creed of the Disabled Man" urged veterans to become "a MAN among MEN in spite of their physical handicap."[45] Throughout the West, social scientists conflated the male body with maleness, believing that "an incomplete version of the former could, without careful training and rehabilitation, destroy one's sense of the latter."[46] Bodily integrity was both a function and a basis for a man's identity. Lieutenant Henry Mock of the FBVE, argued that rather than casting disabled white veterans "upon the scrap heap," the state must "restore them to suitable skilled positions now temporarily occupied by women and colored workers."[47] Mock claimed that the apparent unsuitability of "the unassimilable Chinese, East Indians and Negroes" as sources of additional labor meant that the nation's greatest source of labor lay "in making that which we have [that is, white workers] doubly efficient."[48] Progressives were convinced that "manpower," a gendered and racial vision of natural energies that linked masculinity to physical exertion, needed to be replenished at all costs along strictly defined racial lines.

The Worth of Wounds: Diagnosing and Monetizing Disability at the Bureau of War Risk Insurance

For FBVE officials, the diagnosis of disability coincided with its commoditization. Veterans' disabilities were rendered invisible outside the compensatory cash nexus established by the FBVE. According to the Vocational Rehabilitation Act of 1918, "Persons who had been disabled through their service in the military or naval forces of the United States—whether caused by injury, disease, or aggregation of a previous medical condition—were afforded vocational training and assistance in obtaining gainful employment."[49] Individuals could not enter training until awarded compensation by the Bureau of War Risk Insurance (BWRI). The services and facilities of the Public Health Service (PHS) were used to provide examination, medical care, treatment, and hospitalization for beneficiaries of the War Risk Insurance Act. Initially, the Rehabilitation Act was intended "to divorce compensation from patriotic sacrifice or physical or mental suffering and instead link it to the reimbursement of potential lost future income due to disability and the general disadvantages it caused the veteran in the labor market."[50] Vocational rehabilitation produced a disabled identity that, though rooted in the body, was essentially an aggregate of social relations relative to specific industrial occupations and hierarchies—an identity that was little more than the sum of its parts.

Given that few African American ex-servicemen suffered combat wounds, their experience of vocational rehabilitation differed from that of their white peers. African Americans' initial diagnosis of disability did not occur during convalescence in domestic or overseas field hospitals, but upon submission of their disability claims through the BWRI. In the five years following the end of the war, approximately 930,000 veterans applied for disability benefits. Though applicants were required to indicate their race on their appraisal forms, the FBVE and its successors at the Veterans Bureau did not compile statistics on the filing and rejection of claims based on race. Walter Hickel notes that "racial segregation was integral to the social networks within which disability was diagnosed."[51] One bureau official observed, "Southern representatives who are of course always white, will not, as a matter of principle, forward us all the necessary evidence to complete the Negroes' claim. As a rule, the fact that the claimant is a Negro in their eyes is sufficient evidence that he is not in need of disability assistance."[52] FBVE officials were generally disinclined to provide compensation to a people whose apparent worthlessness informed the adjudication of white entitlements.

Black veterans and their allies in the black press vociferously defended their right to rehabilitation. The *Chicago Defender* speculated that "out of this great world struggle may come industrial and civil freedom . . . certainly the colored soldier who fights side by side with the white American will hardly be begrudged a fair chance when the victorious armies return." Robert R. Moton, Booker T. Washington's successor at Tuskegee, told returning black soldiers, "America is a great laboratory which God is using to show the world how men and women of different races can succeed together." *The Crisis* urged that "industry be extended to the colored American and the same unlimited opportunities to serve in Reconstruction as it did in war times when men of both races 'went over the top' often suffering the most grievous of wounds." Some black veterans conflated their physical wounds with those inflicted by the "prejudices, insults, and duplicity of his white American ally." Within the context of the New Negro ethos, these dual factors had made the black soldier into a "super-soldier and superman" deserving of compensation and the nation's respect.[53]

Contrasted with traditional military pensions, the model developed by the BWRI was not calculated in relation to a veteran's prewar or current occupation. Instead, it measured the average reduction in earning capacity that a veteran with a specific disability was likely to incur in any skilled or unskilled occupation. Since reduction in earning capacity varied with the type and severity of the impairment, disability was expressed as a percentage, representing the deviation between the estimated production of an "average" working body and the residual capacity of a disabled veteran. To help in the calculation of this percentage for "specific injuries of a permanent nature," the BWRI developed a disability-rating schedule with a comprehensive index of amputations, injuries, diseases, and mental disorders. This schedule assigned percentages for each impairment based on its purported effect on the veteran's ability to work. While rates of compensation varied according to marital status and dependents, veterans accorded a 10% rating or more for a temporary disability were entitled to approximately $80 to $120 a month. Those judged to be suffering from total disability, including blindness, multiple amputations, or the "helpless and permanently bedridden," were assessed at a flat rate of $100 a month.[54]

Veterans' benefits were dispensed as inducements for the disabled to redirect their productive energies in unprecedented and profitable ways. Westinghouse's medical director, Charles Lauffer insisted that "in this age of specialization and diversified industry, arms and legs are really incidental, for with mechanical devices, such handicaps are virtually overcome."[55] For ex-servicemen with serious physical impairments, officials stressed the mental acuity needed to perform skilled industrial labor: "While from his neck down a man is worth about $1.50 a day; from his neck up, he may well be worth $100,000 a year."[56] Conversely, appeals to veterans experiencing mental disorder emphasized the need for physical vigor and resistance to fatigue required for modern industrial work. One army official noted, "A man is crippled only to that extent to which he allows his physical handicap to put him down and out. If he ceases to be an economic factor in society—an earning, serving unit—he is a cripple."[57] A postwar managerial elite weaned on industrial management and evolutionary theory framed disability as both an identity and a commodity: a corporeal index of composite labor capital or lack thereof.

FBVE officials believed that their disability-rating schedule democratized veteran entitlements by defining disability in relative terms—as a declension from a productive or "normal" body. However, the ethos of early twentieth century republican manhood dictated that the ideal or normative working body was that of a white man. Socioeconomic dependency and non-citizenship were traits most often associated with as seen as the purview of women, children, and people of color.[58] Both FBVE officials and black veterans understood disability in relation to their respective social relationships and needs as men. In defining disability, both groups "emphasized the structure of local labor markets, racial segregation, and social norms, which assigned men their roles as workers, providers, and citizens."[59] For all veterans, disability denoted more than an individual medical condition or a purely legal entitlement to benefits. Black veterans saw their injuries as marks of patriotic sacrifice that entitled them to veterans' benefits, whereas rehabilitation officials saw them as indelible proof of blacks' pathological inferiority.

Medical models of racial disability drastically undermined African Americans' claims to vocational rehabilitation programs. FBVE doctors contended that the majority of African American veterans were disproportionately afflicted with the hidden wounds of "colored diseases," such as venereal disease and tuberculosis. Local officials—like the FBVE manager of District Five (comprising the Carolinas, Georgia, Florida, and Tennessee)—were suspicious of black veterans' claims, and suggested that "the majority of the disabilities of the southern Negro are traceable to TB and VD which in the majority of cases were judged to have existed in the race as a whole before enlistment."[60] Black veterans vociferously rejected these characterizations with overt appeals to a common manhood. The NAACP urged that "industry be extended to the colored American and the same unlimited opportunities to serve in reconstruction as it did in war times when men of both races 'went over the top' often suffering the most grievous of wounds." Black veterans called on the federal government to honor their wartime sacrifice by providing unfettered access to vocational rehabilitation.[61]

Notwithstanding their race-neutral posture, many FBVE officials recognized that benefits paid to black veterans could potentially undermine white dominance rooted in income distribution, regional labor markets, and citizenship rights. Compensation could amount to several times the thirty dollars a month black agricultural laborers earned on average in the South, and it could more than equal the $500 to $600 annual income of most rural black families. Though modest, these benefits would enable black veterans to temporarily forego poorly paid menial "colored" jobs, and gain a small measure of financial independence.[62] The ensuing financial and educational opportunities accorded black veterans could also work to undermine many of the dubious property and literacy requirements that restricted black voting rights in the South. The FBVE's characterizations of blacks as congenital racial cripples rationalized the economic and social self-interest of southern and national white supremacy.

Fitting the Right Races to the Right Places: Black Veterans and Vocational Training at the FBVE

African American veterans' difficulties in drawing compensation—with its pernicious subtext of black dependency—invariably circumscribed their access to job training. Sympathetic observers linked war to work to argue for blacks' right to rehabilitation: "The Negro soldier left a civil occupation to take up his gun. These recruits were not loiters, but laborers in the larger sense. They were contributing to the nation and the world's work in various ways both skilled and otherwise."[63] In the initial stage of physical reconstruction conducted under the Office of the Army Surgeon General, disability was perceived as a source of entitlement to receive re-training. When the process shifted to vocational rehabilitation, disability was redefined as a chance to remake oneself. For FBVE officials, this represented an unprecedented opportunity for the "crippled Negro race" to work its way to civilized respectability. Vocational rehabilitation connected racial uplift to patriarchy through its efforts to "not only make the Negro a better workman, but also teach him to build a better home and live a more ideal life." FBVE policy was motivated by "the idea to elevate the economic status of these [black] men sufficiently to enable their children to attend school and their women to give more time to the moral and hygienic development of the home."[64] Across the color line, the ethic of rehabilitation equated fit bodies with healthy homes.

Black veterans' wounds were a form of social currency that supposedly entitled them to vocational rehabilitation. Initially, the state appeared willing to let blacks perform this transaction in the name of national efficiency. One bureau employee remarked, "The worth of our Negro veteran rests not in his racial nature but in his ability to work himself into an efficient and productive worker."[65] Another noted that "if the Negro is coming into our lives to stay—and we need him—we should recognize the fact that he is perfectly capable of profiting under vocational instruction and becoming a worker worthy of the name."[66] The Crisis was confident that the FBVE would "surely undertake—regardless of race—the training of a disabled soldier for a new occupation or retraining to better fit him for his former occupation." Most FBVE officials were not so accommodating. Frederic Keough countered, "The fact that a man is a disabled soldier or sailor is not enough to place him in any systematic manufacturing plant. He must be productive." Most importantly, Keough concluded that vocational rehabilitation "must terminate in an economic advantage to the community."[67] However, amid what historian Matt Guterl describes as a "southernization" of postwar national race relations, appeals to rehabilitation's "communal advantages" drastically limited the kinds of work African Americans could do and where they could do it.[68]

In summer 1918, the FBVE established the Rehabilitation Division to undertake the vocational education and placement of veterans. This division maintained three types of offices: the Central Office in Washington, D.C., fourteen district offices composed of two or more states, and one hundred or more local offices. Each district office was headed by a district vocational officer who presided over two or more assistant district vocational officers. One of these officers was responsible for training supervision in the local offices, and another for industrial relations and employment aid.[69] Dr. J. R. Crossland, a self described "fearless defender of the claims of the Negro," was appointed by the FBVE as a "special expert on Negro affairs."[70] Though some in the black community saw Crossland's post as misguided and little more than ceremonial, he was a key mediator between black ex-servicemen and the often-Byzantine workings of the FBVE.[71]

FBVE officials used the model of disability laid out in the BWRI disability schedule to train injured veterans in a variety of occupations. According to this model, a man was not "handicapped" by his physical condition but only by the specific limitation that condition placed on his employment. Although "one may find himself handicapped in one occupation he may not be in another."[72] Fears that disabled veterans, and the disabled in general, would become burdens on society led to repeated calls that integration, not segregation, was essential for any program of rehabilitation to be truly effective. Officials believed that "there is a danger inherent in the reservation of specific employment for disabled men. It makes a special class of cripples; employments reserved for them cannot fail to become characterized as subnormal occupations."[73] Having long been segregated into menial "subnormal occupations," African American veterans now found them trapped in a double jeopardy of race and disability.

The FBVE's "one size fits all" policy of reintegrating veterans into the general work force collapsed in the summer of 1921. Amid mounting allegations of corruption and inefficiency, the FBVE was absorbed into the Veterans Bureau in August 1921. Consequently, all federal officers were eliminated and complete authority for determining the eligibility of ex-service men for training was delegated to the district offices. In spring 1923, Crossland protested to bureau officials over cuts to his already limited staff, and the ominous disconnection of his office telephone. He argued that as the lone federal representative of black veterans, it was essential he "keep in close contact with local businesses, community centers, the Red Cross and various other rehabilitation institutions." Crossland's pleas went unanswered.[74]

The decentralization of the FBVE allowed local mores of race, labor, and disability to emerge as the dominant model in the bureaus' day-to-day operations. Given that four out of five black veterans still lived south of the Mason-Dixon Line, southerners exercised a tremendous influence over FBVE policy. During the mass exodus of blacks to the North, east coast elites increasingly deferred to southern "racial expertise" to deal with the sudden "Negro problem" in their midst. Medical models of rehabilitation allowed southern FBVE officials to couch their traditional animus towards blacks in the more palatable rhetoric of scientific racism. Noting tubercular blacks' apparent inability to undertake "strenuous vocational rehabilitation," officials in the FBVE central office in Washington, D.C. concurred with their southern counterparts that "tuberculosis is in the colored race as a whole."[75] Consequently, any attempts to retrain tubercular blacks for industrial labor were dismissed as not only foolish but as contrary to the very laws of science and nature.

Vocational rehabilitation in the South was a key mechanism of racial labor division and control. Biological rationales for black inferiority bolstered the socioeconomic imperatives and hierarchies of Jim Crow. And though these rationalizations were undeniably racist, they were not irrational in the context of contemporary political economy. Blacks were indeed indispensable to the southern rural economy. Blacks comprised 48% of all southerners engaged in agriculture; cultivated two-thirds of the region's land; owned or rented 41 million acres of farm land worth approximately one billion dollars; and tilled some 60 million more as laborers. Southern elites prevailed upon the FBVE to provide the "Negro with the kind of education that he needs and demands, namely, vocational agricultural education."[76] From November 1917 to November 1918, federal funds subsidized the creation of thirty-nine vocational education schools for black veterans and workers. FBVE agent H. O. Sargent proudly noted that "in these schools little attention is given to preparing students for college" and that "classes were directed solely to the productive field of farming," thereby keeping the Negro in his proper place.[77]

FBVE officials drew on established forms of black industrial education in the design of vocational rehabilitation programs. Although Hampton-style education in the semiskilled "mechanical arts" had long been a staple of the prewar South, federal officials argued that there were too few teachers or institutions to facilitate the process. To fill this gap, the FBVE and its successors at the Veterans Bureau committed to the establishment of new vocational schools. These schools were divided into three types: urban schools for non-resident students, training centers for resident students, and reconstruction centers in Public Health Service hospitals. The curriculum was designed to meet the needs of the local white community. For black pupils—who generally functioned at a fifth to eighth grade level—at least three hours a day were devoted to vocational agriculture, half of which was spent on supervised study and laboratory work, while the other half was devoted to its practical application. Each student carried out a project that generally involved the raising of some farm crop or farm animals, usually at the home of the student. White officials never hesitated to note examples of blacks' supposed affinity for farming. When two out of three disabled black veterans in training at Tuskegee "looked out over all the vocations open to them and chose the land" bureau, officials cited this as proof that "the call of the wild isn't any stronger than the call of the land" and commended the two for their "courage" in answering to the latter. For Southern FBVE officials, the Negro agricultural worker was a necessary and natural entity.[78]

Violence—both real and imagined—sustained Jim Crow and invariably colored blacks' experiences of rehabilitation. When FBVE officials assigned Oscar Woods to undertake tailoring training in Georgia despite his extensive background in auto mechanics, he protested to special Negro Advisor J. R. Crossland: "I would prefer to pass through Alabama in an aeroplane, driving fast at that, of course . . . on account I hear they have the KKK in Georgia and Alabama and I know how they hate working niggers." Woods fearfully insisted that if he was sent to Alabama, "when you bring me back, I won't need no meal ticket. My flag-draped casket will be enough."[79] This combination of regional racial labor mores and hierarchies, along with the constant threat of violence, forced many African American veterans to confront the brutal and practical limits of vocational rehabilitation.

Characterizations of blacks as cripples—the conflation of racial and physical impairment—were by no means confined to the South. Northern FBVE officials also maintained a proscribed list of semi- and unskilled trades for the minority of African American veterans who qualified for vocational rehabilitation. Throughout the nation, African American veterans were disproportionately placed in training for shoemaking and tailoring, with auto mechanics—the sole mechanical occupation—ranking a distant third. If blacks selected trades that did not appear on the approved list, bureau officials were instructed to keep them out of training indefinitely. When Mack Hudson of Philadelphia reported to his Local FBVE Office upon being certified for training with an undisclosed injury, he requested auto mechanics but was offered shoe repair. When Hudson refused, he was told that the board "could do nothing more for him."[80] J. R. Crossland protested, "It appears from the complaints which constantly pour into my office that several districts believe there are certain occupations in which they cannot afford colored men. And it is not always true that the necessary facilities are not present and readily available. The whole thing is working a great injustice upon the men of my racial group."[81]

Notwithstanding these restrictions, the majority of disabled African American veterans—emboldened by their wartime service and perhaps accustomed to the occupational segregation of Jim Crow—complied with their vocational placements. One official observed that most men went "into training at the trades they are given not with the view to being rehabilitated, or even ever working at the particular trade, but simply to draw the training pay for the allotted time." Like all veterans, African Americans placed more importance on their financial benefits than they did on the actual practices of physical rehabilitation. Bureau officials constantly complained that African American ex-servicemen were "lazy" in not taking up vocational training and should therefore be denied compensation. Similar charges of negligence were leveled against veterans of all backgrounds, but only African Americans were continually and disproportionally denied compensation and job training by the War Risk Insurance Act and the FBVE.[82]

To Make the Negro Anew: Race and Health at Tuskegee Veterans Hospital

Debates surrounding the racial dynamics of veterans' policy culminated in the development and operation of United States Veterans Hospital No. 91 at Tuskegee, Alabama. Founded in 1923, the hospital's guiding principle was on institutionalization, rather than rehabilitation. Hospital staff were less concerned with "making the Negro anew" than managing their patients' disabilities. Palliative care was the norm, and segregation—occupational, residential, or social—became the rule of the day. In contrast, black veterans believed residence at Veterans Hospital No. 91 endowed their injuries with the stamp of federal authority that entitled them to the compensation needed for vocational rehabilitation.[83] And indeed, contrary to the army's rigorous enforcement of Jim Crow, racial segregation did not officially extend to rehabilitation hospitals. FBVE officials' commitment to efficiency and standardization, and blacks' real and imagined absence from the ranks of the "deserving disabled," accounted for a seemingly integrationist and race-neutral policy. However, the Surgeon General's Office repeatedly claimed that it had "no intention . . . to settle the so-called Race Question," which paradoxically made it willfully blind to the day-to-day dynamics of race relations.[84] Ultimately, the lack of de facto segregation did not preclude the practice of de jure racial segregation throughout the veterans hospital system.

Hospital authorities throughout the nation often refused to hospitalize black veterans in integrated institutions for fear that their mere presence would undermine patient morale and disrupt the local community. Especially troubling was the specter of "race mixing," in which white female nurses or physiotherapists could find themselves in proximity to the crippled, diseased, and half clothed bodies of African American veterans.[85] For those blacks "fortunate" enough to be admitted to Public Health Service (PHS) hospitals, treatment was typically administered in inferior segregated facilities. One veteran, Isaac Webb described his experience in a letter to *The Crisis*:

> I am also one of the boys who volunteered in 1917 for services "over there" and I have spent six months in hospitals for the disabled. . . . At Mobile, I was handed my food out of a window, forbidden to use the front of the hospital to enter my ward, given no medical attention, and forced to use the same toilet facilities fellows in advanced stages of syphilis and gonorrhea used.

Consequently, many African American veterans refused to seek treatment at PHS facilities, driving black hospitalization rates down to almost 50 to 80% below that of whites.[86]

Tuskegee Veterans Hospital grew from the efforts of the Consultations on Hospitalization—or the aptly named White Committee—that convened in the spring of 1921 under Secretary of the Treasury, Andrew Mellon. Secretary Mellon enlisted a committee of medical experts who labored for two years to create a veterans hospital system that was "not only national, but rational" in scope. Dr. John Farris, head of the Red Cross Institute for the Blind and Disabled, echoed this commitment to dispassionate rationality. He argued that rehabilitation hospitals "should be the nursery of new hopes and ambitions, and not a Bridge of Sighs."[87] These new facilities would be organized and financed by the federal government and initially restricted to veterans with combat service-related diseases and injuries. The committee was assisted by an advisory group that included representatives from the PHS, the National Committee for Mental Hygiene, National Home for Disabled Volunteer Soldiers, and the National Tuberculosis Association. The committee's final report recommended that a separate national hospital for black veterans be established at Tuskegee, Alabama. Though the original legislation that established the hospital system did not mention separate facilities for black veterans, the committee noted early in its deliberations that "one of the great American problems—that of race—obtruded itself more and more."[88]

Debates over whether to employ a white, black, or integrated staff spoke to broader issues of professional expertise, racial segregation, and the right to define disability. The black-focused National Medical Association and the NAACP aggressively lobbied the federal government to employ black staff at Tuskegee to maintain the school's longstanding "commitment to race betterment." Along with providing much needed jobs for African American physicians and nurses, it was also felt that blacks would provide better care to patients due to their mutual "racial affinity." Dr. J. F. Lane of Lane College remarked, "If we cannot serve our own people, where shall we work and whom are we to serve?"[89] The White Committee rejected Lane's petition due to a lack of black orthopedic surgeons and a general aversion to recognizing any form of African American professional expertise. In the end, the committee elected to staff the hospital with all-white personnel.[90] Much like their forebears, disabled black veterans at Tuskegee were often viewed as objects of pity, solely dependent on the paternalist largesse of whites.

Demands for an all-black staff increased in the wake of the hospital's official dedication. In May 1923, General Frank T. Hines, Administrator of Veterans Affairs in Washington, D.C., asked Tuskegee's Dr. Robert R. Moton whether he thought it advisable to staff the hospital with black doctors. Moton replied, "Inasmuch as all the patients will be Negroes and since Negro physicians are not allowed at present to practice in any large hospitals, it would be fair to give them this opportunity."[91] Tuskegee whites were infuriated at the prospect of "colored doctors," and the local Ku Klux Klan (KKK) staged a number of dramatic and violent protests on the hospital grounds in the summer of 1923. The Klan denounced the presence of "carpet-bagging negro professionals" and demanded that the hospital maintain an all-white staff, even though this contravened a state law prohibiting white medical personnel—specifically female nurses—from treating blacks. Klansmen took the remarkable step of privileging white medical expertise over the alleged preservation of white womanhood, stating, "We do not want any niggers in this state who we cannot control."[92] While many in Washington were more than sympathetic to the Klan, most officials could not countenance attacks on federal property. The Republican administrations of Warren Harding and Calvin Coolidge reluctantly agreed to turn the hospital administration over to African Americans, to curry favor with the growing bloc of urban black voters in the North.[93] Tuskegee Hospital's white director Dr. Robert Stanley was quickly replaced with Dr. John Ward, a leading figure in black medical circles who had served in France.[94] Ward arrived at Tuskegee in July 1924 and directed his staff to maintain their focus on providing patients "sympathetic aid and comfort." Ward also opposed occupational therapy, suggesting that "veterans would have little use [it], given the social strictures placed on their advancement"—further demonstrating postwar rehabilitation as a function of labor economy.[95]

The types of disabilities and diseases—primarily tuberculosis (TB)—found in the patients at Tuskegee Hospital largely informed its focus on paternalistic or palliative care. Though tuberculosis had exacted a debilitating toll on the wartime army as a whole, black troops suffered from the "white plague" at a rate nearly double that of whites. A 1921 study by the Army Surgeon General revealed that the highest admission rates for TB were found in black troops whether in Europe or stateside or solely white troops in Europe. Citing earlier environmental or climatic theories of racial difference, officials concluded that while the seasoned white soldier experienced "a marked advantage both as to the admission and death rates, the effect of seasoning on the colored soldier was much less marked, and indeed under the conditions he was called upon to face in Europe his admission rate was much higher than that of the relatively untrained colored men in this country."[96] Uprooted from their "natural" southern habitat and transported to the foreign climes of northern France, blacks were seemingly unable to ward off these new and more aggressive strains of the tubercular bacillus. Tuskegee Hospital No. 91 had accommodations for approximately three hundred veterans suffering from tuberculosis and just over three hundred spaces for those afflicted with neuropsychiatric disorders. For roughly the first two years of the hospital's operation, approximately 60 to 70% of its patients were listed as tubercular.

Analysis of monthly reports from the resident American Red Cross Director at Tuskegee provides substantive insight into the daily workings of Veterans Hospital No. 91 and the racial politics of disease etiology. Initially, the Red Cross was intended to secure patients' social histories, particularly in the neuropsychiatric service. However, as the hospital's emphases on social welfare and non-surgical/curative methods of care increased, so, too, did the power of the Red Cross. These records—spanning the period from the hospital's opening in April 1923 through to December 1926—chart the hospital's shift from an all-white to an all-black staff and the consistent focus on palliative rather than rehabilitative care. Though black physicians, nurses, and various hospital officials did tend to provide patients with more "sympathetic forms of care," they rarely, if ever, questioned the directives of the Veterans Bureau to manage rather than treat their patients. Occupational therapy and vocational rehabilitation were in short supply at Tuskegee throughout the 1920s. Ex-servicemen received only the most rudimentary vocational training in agriculture and the mechanical arts. FBVE officials concurred that tubercular blacks must be treated in the "healthful and restful climate to which they were accustomed, much like that at Tuskegee."[97] In late 1923, the FBVE, in conjunction with the National Tuberculosis Association, published a manual for vocational advisors detailing suitable occupations and vocations for tubercular veterans. The manual was the culmination of a two-year study on the negative physiological effects of tuberculosis on an individual's labor capacity. The study's authors were eager to incorporate the work of Jules Amar, Director of the Research Laboratory of Industrial Labor in Paris, who had developed a new method for measuring the lung capacity of individuals suffering from pulmonary tuberculosis. FBVE officials recommended that the U.S. government petition Amar to accept an advisory position at the Veterans Bureau to pursue studies on tubercular laborers. Given Amar's work on colonial African troops, the Veterans Bureau expressed interest in commissioning a study at Tuskegee on African American veterans. Despite the failure to implement these studies, their very commission represented an intriguing attempt by FBVE and Veterans Bureaus officials to decipher the racial etiology within a broader transnational economy of racial labor fitness.

Red Cross records also revealed the modest ways in which black veterans attempted to counter characterizations of them as fundamentally disabled. In May 1924, just two months before turning over the hospital to an all-black staff, an occupational therapy department opened on the hospital's grounds. Red Cross director E. M. Murray reported that "many of the men were loud in their expressions of delight in its beginning" and that the "work will do much toward abating any spirit of restlessness that might have existed among the men. Indeed it has already reduced the innumerable requests made upon this office for occupational supplies." Through their persistent demands for occupational/vocational supplies, black veterans actively resisted the paternalistic regime of "enforced idleness" that had characterized their treatment.[98] African American veterans saw such treatment as an affront to their identities as soldiers, workers, and men. Evelyn Z. Phelps, Director of the Red Cross

Service at Tuskegee, remarked, "The patients rebel at the long rest hours, and many complaints are heard. They are kept in bed about twenty hours out of the twenty four, and feel keenly the lack of reading material, occupational therapy, vocational training and teachers."[99] Despite the introduction of an occupational therapy department and an all-black staff, Tuskegee continued to maintain a palliative rather than a rehabilitative focus, committed to the warehousing of broken black bodies.

Hospital officials defined broken black bodies as symptomatic of broken black homes. Reviewing the patient files of Mr. Tuttle Duke—a divorcee and "rather foppish fellow"—a Red Cross official remarked that his "enfeebled physical state as a tubercular" was matched "only by his predilection for vice so characteristic for a Negro of his type."[100] In their monthly reports, hospital officials routinely derided "deformed and malformed" patients as "unfit guardians" and "absentee fathers and husbands."[101] The diseased and deformed bodies of black veterans were read as evidence of their debased familial values, and a threat to the broader social fabric.[102] One staff member noted that patients should be denied the opportunity to return to their families, even if "many feel that they can take the treatment as well at home and worry needlessly about their families, for if the truth were known, many of the families are far better off when the patient is away." In early 1925, the new Red Cross director E. M. Murray cited the increasing "tendency of patients bringing their families to Tuskegee as something which we are endeavoring to prevent as much as possible."[103] Hospital officials traded on prevailing theories of African American degeneracy as social contagion to rationalize their restrictive forms of care.[104]

Tuskegee Hospital's emphasis on palliative care reduced its patients to little more than wards of the state. At its peak in 1923, the FBVE was active in rehabilitating approximately 2,500 African American ex-service men, with an additional 1,500 in hospitals—the majority of whom were housed at Tuskegee. Shortly before the hospital's shift to an all-black staff, Evelyn Z. Phelps of the Red Cross remarked, "It is our experience that the Trust Companies make the best guardians and that relatives of the patients make the poorest." This was because "some of the families of the Negro patients are wholly ignorant and illiterate and pay no attention to the letters which we write them requesting aid for their kin." Phelps observed that black families mistakenly "have the idea that the government is caring for the man and that every necessity is provided."[105] Officials' disgust at poor blacks' apparent lack of familial commitments reveals the chasm separating black and white understandings of veterans' entitlements under Jim Crow. To the degree that black veterans and their families often viewed hospital care as their patriotic right and entitlement, whites—who were far more likely to receive care—nevertheless saw it as a privilege, dispensed by a beneficent state to an inferior people.

Tuskegee's paternalist ethos also exposed intra-racial class tensions. Staff drawn from the African American elite—what W. E. B. DuBois deemed the "talented tenth"—worried that the mental and physical deficiencies of the "lower classes of Negro patients" would impede the race's progress to respectability.[106] From 1923 to 1927, Red Cross workers placed close to two hundred ex-servicemen in local trusts such as the Bank of Tuskegee.[107] Hospital workers also took steps to confine some of the severely mentally and physically disabled to jails and insane asylums, effectively criminalizing perceived racial disabilities. An especially egregious example occurred when the Red Cross and Veterans Bureau sent thirty tubercular African American veterans to the Central State Hospital for the Criminal Insane in Nashville, on the spurious pretext that "their condition could be attributed to a uniquely racial mental affliction."[108] Tuskegee officials, whether black or white, eschewed vocational rehabilitation in favor of the various disciplinary institutions of the postwar state.

Postwar policies and practices of rehabilitation evoked antebellum techniques of the discipline of black bodies under state and federal auspicious. Yet rehabilitation was a decidedly modern phenomenon "born as a Progressive Era ideal that took shape as a military medical specialty, and eventually became a societal norm in the civilian sector."[109] The efforts of rehabilitationists to delineate the "deserving" from the "undeserving" disabled ultimately framed citizenship in corporeal terms. This, in turn, drastically circumscribed opportunities for injured black veterans whose physiognomy had historically marked them as defective—prisoners to their irredeemably primitive bodies.[110] FBVE officials conflated blackness, disability, and dependence—the antithesis of republican citizenship—to consign black veterans to the margins of the interwar labor economy.

Rehabilitation—through its unique ability to link the public and the private sectors around issues of veterans' provisions—redefined race, labor, and citizenship in postwar America. Henri-Jacques Stiker reminds us that models of rehabilitation that emerged from the Great War reconfigured disability from a curative condition tied to notions of "removal and individual health," to a "lack to be overcome," or a "deficiency to be remedied" through various legislative and institutional channels.[111] Through these new forms of rehabilitation, race became a key marker of difference or deficiency to be overcome, in the real and symbolic postwar economy of disease and disability. Yet, given rehabilitationists' oft-cited beliefs in African American inferiority, even limited forms of rehabilitation appeared to run counter to the core tenet of modern veterans care: autonomy before charity. Elites in both the public and private spheres worried whether the state should actively enable "the worst dependent traits of the colored race," in a misguided and costly attempt to salvage the unsalvageable. Rather, the state should simply "cut the colored man loose" and let the dysgenic natures of war and industry finally do away with the burdensome "Negro problem."[112]

FBVE policies, while profoundly racialist, were not driven by explicit racial prejudice. In fact, race hatred was quite beside the point for its architects. The successful management of the war effort emboldened social scientists to rationalize human systems along biological lines.[113] Rehabilitationists saw themselves as self-appointed guardians of evolutionary processes—weeding the fit from the unfit and pruning society of its most undesirable elements. FBVE officials rationalized these practices as necessitated by wartime imperatives and not by racial animus. Managerial elites believed that limiting African American veterans' compensation, consigning them to menial occupations, and isolating them in segregated health care facilities were all necessary interventions in the processes of social evolution.[114]

Official claims to scientific objectivity were belied by the use of widespread state coercion. Unlike their European counterparts, American rehabilitation policies were deemed mandatory in the hopes that more men restored to the workplace would result in reducing the costs of federal disbursements. But implicit in the practice of rehabilitation—the drive for "aggressive normalization"—was a regulatory impulse to contain bodily and social difference, so readily embodied in the figure of "the Negro." Through these processes, evolutionary science combined with scientific management to reconcile racial form with labor function. Prewar ideas of race as a multihued phenomenon—a cacophony of races—gave way to a stark postwar biracialism of black versus white. These new forms of racial knowledge linked race and labor fitness to color and the body: the healthy, normative, white working body was juxtaposed with the degenerate, abnormal, black working body. The wounds of black veterans were seen not as badges of patriotic honor, but as the stigmata of atavistic agents threatening to poison the body politic from within. Vocational rehabilitation was ultimately a discursive and legislative tool deployed by postwar managerial elites to come to terms with an emerging black proletariat. New methods of racial labor division and control were required as the Negro moved from the farm to the city. For even the most liberal-minded reformer, the goal was never "to make the Negro anew," but to contain, or at best alleviate, the far-reaching and dangerous socioeconomic effects of his inevitable slide into degeneracy.

FBVE models of vocational rehabilitation perpetuated and institutionalized prevailing notions of the Negro as diseased, deviant, and congenitally unfit for modern industrial life. But these ideas did not remain constant over time. Rather, they were repeatedly challenged by the exigencies of war, migration, urbanization, and African Americans themselves, in ways that eventually helped sever race from biology. But this gradual shift from biological to cultural understandings of racial difference did little to change the fortunes of the average African American veteran who remained on the margins of the postwar labor economy. For the ideological and institutional architects of postwar white supremacy, the crippled and deformed body of the disabled Negro veteran was a harbinger of America's prospective racial decline: the canary in the nation's evolutionary mineshaft.[115]

5

A New Negro Type

The National Research Council and the Production of Racial Expertise in Postwar America, 1919–1929

"The race problem is rather a difficult one because we do not know exactly what Negroes are or what the Negro race is."
—Dr. Joseph Peterson, National Research Council (1928)

Notwithstanding the conflict's gruesome human toll, the war also shook progressives' faith in their ability to effect historical change through rational reform. One economist declared, "It would perhaps be an exaggeration to say that the European war has rendered every text in the social sciences out of date . . . but not much of an exaggeration."[1] As latecomers to the conflict and an ocean removed, Americans managed to avoid the war's most destructive excess yet as active participants in the discourse of transatlantic reform, they were unable to escape the pervading crisis of faith that consumed Western progressives. Progressives' notion of progress was linear, linking the past, present, and future through the medium of empirical inquiry; to know the past was to shape the present and the future. Progress was consequently imbued with a sense of perpetual motion, an inevitability that could only be undermined through inattention—willful or otherwise—to the laws of industrial evolution which progressives had long conceived of in explicitly racial and corporeal terms. The war had accelerated a process of "racial impoverishment" and unleashed an insidious atavism embodied in "the colored hordes" that now threatened to penetrate the "outer and inner dikes" of the white world—laying siege to the very future of western civilization.[2]

The successful management of the war effort provided social scientists with the impetus to rationalize humans and human networks through theoretical models analogous to those in the natural sciences—effectively creating an organic model of political economy along racial lines. Surveying the ravaged postwar landscape for clear answers to the era's seemingly insurmountable socioeconomic, political, and cultural problems, progressives reconfigured social processes as a function of biology or, more specifically, race. The war further convinced progressives that "biological diseases required biological cures."[3] Wartime imperatives produced racial and labor taxonomies through the draft and rehabilitation to combat the human "waste of war" and purge the body politic of its most inefficient (and therefore undesirable) elements. Despite the war's horrors, the conflict had provided "extraordinary opportunities for the study of different races and their various physical and mental aptitudes."[4] For many postwar observers, taking stock of the war's dysgenic effects was paramount. Avowed white supremacists such as Lothrop Stoddard saw the conflict as a white civil war that portended the fall of Western society. For many, this decline had begun over a decade prior with Japan's shocking victory in the 1904 to 1905 Russo-Japanese War, which vaulted the Asian nation into the ranks of global imperial powers.[5] In the intervening years, new technologies had increased the flow of global migration and capital, spurring the advance of the colored races from east to west.

Fears of a global rising tide of color inundating white civilization were constitutive with the development of the postwar U.S. state. Drawing on wartime testing, the authority of the postwar state was predicated on its ability to maintain the racial and corporeal integrity of the republican body politic. These new forms of eugenic statism informed a wide variety of public initiatives from federal funding of the National Research Council (NRC) to hyper-restrictive immigration legislation. Policymakers in Washington drew on their respective wartime experiences in concluding that the nation's productive capacity and democratic health required inviolable borders and bodies. Even prior to the armistice, the biological effects of the "war to end all wars" were already clear: the best of white manhood had been slaughtered in the trenches, leaving the unfit behind to breed.

Perhaps most troubling to white elites were the masses of colored laborers and soldiers who had traveled to the various metropolises of empire—New York, Chicago, Paris, Berlin, and London—to labor at the work of war in the service of their colonial masters, many of whom now intended to stay. For Stoddard, "colored triumphs of arms are less to be dreaded than more enduring conquests like migration which would swamp whole populations and turn countries now white into colored man's lands irretrievably lost to the white world." Though "these ominous possibilities existed even before 1914, the war had rendered them much more probable," due in large part to the white world's unbridled avarice.[6] Professor Warren S. Thompson of Cornell University remarked, "One may look upon the dying out of those who worship the God of Mammon as nature's kindly provision for ridding the world of the overambitious, egotistic elements who have missed the true goal of living." In abrogating their duty to propagate the "best of the breed," whites now faced submersion by the world's colored majorities. The language of inundation was also invoked by the likes of Col. William J. Simmons who sought to revive the Ku Klux Klan in 1915 for the purposes of maintaining "Anglo-Saxon civilization on the American continent from submergence due to the encroachment and invasion of alien people of whatever clime or color."[7] Stateside, this menace was embodied in the New Negro type that, according to some wartime testing, had emerged from the war reinvigorated and poised to reshape the very boundaries of race in America.

Postwar social scientists with the NRC built on wartime studies to develop, sustain, and institutionalize a working body of knowledge on the Negro and black labor. The expansion and articulation of state power evinced in the production of racial labor knowledge would prove to be one of the conflicts enduring legacies to American social thought. Specifically, the idea that race was a knowable, quantifiable entity, an object of inquiry, and a problem to be solved through state auspicious. Efforts to chart the effects of migration, urbanization, and race mixing on an ostensibly new Negro type, which had emerged from the war, produced new forms of knowledge on race—the Negro as embodied pathology—which would inform the discourse of American race relations for the foreseeable future. Indeed, the institutionalization of race and racialized peoples as "problem" was perhaps the most significant consequence of wartime testing. Formed in December 1916, the NRC was empowered "to stimulate research and apply the knowledge to the public welfare; to plan comprehensive researches and to minimize duplication by cooperation; and to gather up and render available existing scientific and technical knowledge" to inform public policy.[8] For the philosopher Alain Locke, the relative "newness" of the New Negro was only "because the Old Negro had long become more of a myth than a man," a "creature of moral debate and historical controversy." Gone was the sturdy yeoman Negro farmer of the Bookerite imagination, dutifully "casting his buckets down" in his "natural" southern homeland.[9] The exigencies of war, migration, and shifting labor processes had transformed the Negro into a mobile, national, yet increasingly elusive figure, despite the era's mania for quantification.

Postwar production of new modes of knowledge on race and the laboring body irrevocably linked race making to state making. Michael Adas argues that scientism—a melding of the "pure" and "social" sciences—informed the efforts of postwar social scientists to establish their authority in drafting public policy and to win funding for their endeavors.[10] Knowledge forged in this public/private nexus was stamped with the imprimatur of "expertise" a quintessentially modern form of professionalized knowledge. The language of "expertise" required that the social sciences be made "as relevant to modern industry as the chemist, engineer or geologist." Sociologist A. E. Jenks cited the need for applied racial labor knowledge noting, "of what value were material gains if civilization itself is to fail?"[11] Determined to influence public policy, NRC officials sought to create a critical mass of expert knowledge that "could assist the American nation and race toward her goal of developing civilization" while simultaneously enhancing the authority of the state.[12]

Conflating categories of race and nation reinforced progressives' belief that delineating the physiological and political contours of the body politic was paramount to national efficiency. NRC officials sought to extend and enhance the states' role in facilitating taxonomies of race and labor fitness as a key organizing principles of the postwar labor economy especially as it related to issues of immigration and migration. Committees on "Migration," "Race Characters/Racial Differences," "The American Negro" and "Anthropology and Psychology" were charged with the collection and dissemination of all the available existing scientific and technical knowledge on race, racial difference and their relation to labor. Before the war, Robert Park had noted that "the simplest problems . . . are world problems, the problems of the contacts and the frictions and the interactions of nations and races."[13] Members of the NRC linked proposals for studies of domestic black migration to a larger global project of racial knowledge production to curry favor and funding from federal officials.[14] Forging the color line at home was essential to stemming the rising tide of color abroad.

The war changed perceptions of black labor fitness in three significant ways. First, wartime migration and military service definitively established the Negro as a factor in industrial civilization. Secondly, rapid migration and urbanization were seemingly affecting mental and physical changes on the Negro—reshaping the corporeal and social contours of blackness in the social scientific imagination. Finally, national efficiency was reconstituted as a function of racial integrity—delineating what bodies could and should do which kinds of work. Socioeconomic, political, and legislative efforts to disentangle race from nation, such as the 1924 Immigration Restriction Act, led to the prewar near-white "races" of Europe being subsumed into an all-encompassing Caucasian archetype. The postwar emergence of the Caucasian as a racial category reduced the prewar multiplicities of white and near-white races to the singular tones of the Caucasian. Race subsequently became a synonym for an increasingly elusive blackness. For better or for worse, to talk of race was to almost exclusively speak of blacks. But as Charles Johnson observed, such a process was fraught with ambiguity, especially in regards to contemporary labor economy:

The Negro worker can no more become a fixed racial idea than can the white worker. Conceived in terms either of capacity or opportunity, the employment of Negroes gives rise to the most perplexing paradoxes. If it is a question of what a Negro is mentally and physically able to do, there are as many affirmations of competence as denials of it.[15]

Though anthropometrical testing had found the "uninfected Negro to be a constitutionally better physiological machine than his white counterpart," skeptics pointed to blacks' woeful performance on the Army IQ tests as evidence of their infantile inability to reconcile the mind and body divide. Even military officials sympathetic to "the Negro's plight" cautioned, "Races develop slowly! A few years ago these men were slaves in the fields and a few years before that they were children in the jungles of Africa. They are children still."[16] Wartime testing framed blacks' innate savagery as a source of their physiological vitality as well as with the mental deficiencies that ultimately undermined said vigor.

Social scientific prewar iterations of the "Negro as problem" persisted into the postwar era. But wartime testing had raised more questions than answers as to just what exactly made a Negro a Negro. Postwar fascination with race mixing and the figure of the mulatto revealed deep anxieties regarding the physiological and social contours of race in America. For Johnson, the "Negro Problem" was "only now in evolution" on a national and global scale.[17] According to DuBois, the war "had given the Negro his chance to widen his narrow foothold on life, a slightly better opportunity to make his way in the industrial world of America."[18] NRC officials concurred, noting that "given that the Negro and his descendants represent a large part of the total population and were now an increasing factor in industrial life, investigation of this group is particularly important, both from a theoretical and practical point of view."[19] To mine the full potential of black labor, it was necessary to study "the Negro's respective strength, endurance, nervous reactions and instinctive responses to stimuli in comparison with those of the white [or Indian] race to discern his capacity for profiting by vocational training." Anthropologist A. E. Jenks remarked that "just as breeds of animals differ, likewise, do breeds (or so-called races) of people differ." For Jenks, the survival of American civilization depended on "the races of men who are to work and breed the nation's future generations."[20] Positing labor economy in stark corporeal terms cast eugenics as an imperative of national health. NRC research on race and its relation to labor transformed prior understandings of blackness as something shaping labor processes and hierarchies—blacks' innate traits and tendencies pushed them towards certain types of work—to one in which form preceded function whereby Negroes were as Negroes did.

NRC officials developed an intricate form of social anthropology, the "science of race and culture," to investigate the "Negro problem" in postwar labor economy. Wartime and postwar testing of blacks were attempts to rectify blacks' previous absence from anthropological discourse.[21] For much of the late nineteenth century, anthropology had focused on the study of what Lee Baker refers to as "out of the way peoples," such as American Indians or indigenous colonial subjects from the American West to the Philippines.[22] Prewar anthropologists paid less attention to "in the way peoples," such as blacks and immigrants—generally ceding their study to the sociological gaze. R. J. Terry, the future chairman of the Committee on the American Negro, claimed, "the Negro Problem is certainly no less complex than the Indian problem and we should have to anticipate the same kind of valuable . . . but fragmentary work being done by the Bureau of American Ethnology."[23] Whereas the Indian was characterized as a hostage to a degraded culture, the Negro was captive to a depraved biology. Franz Boas was one of the few early twentieth century anthropologists to work on the "Negro Problem"—publishing his first article on the subject "The Negro and the Demands of Modern Life: Ethnic and Anatomical Considerations," in the fall 1905 volume of *Charities*, a leading reform organ. Black intellectuals immediately recognized Boas as a key ally in the fight against entrenched racial hereditarianism—his culturist methodologies and his immigrant Jewish roots marked him as a fellow outsider in a discipline and academy dominated by white Anglo Saxon Protestants. In early 1906, DuBois wrote to Boas, inviting him to Atlanta University to speak at an annual conference on the Negro in American Life. Boas's commencement address on the cultural basis of racial behavior and the historic significance of Africa refuted prevailing theories of blacks' biological deficiencies. Linking culture to race, Boas sought to reorient anthropology toward the study of blacks while producing a systemic critique of its white supremacist epistemologies.[24]

When anthropologists eventually did turn to the study of the Negro, they favored three methodological approaches. The first was folklore predicated on the search for the continuity of African culture in the Americas. The second approach foregrounded the nexus of class and race to measure the cultural, social, and psychological toll of racism on black Americans as a whole. The final approach involved physical or biological anthropology that had driven much of wartime anthropometry and now informed the work of the respective committees on migration, race characters, and the American Negro. Advocates of the New Negro Movement used anthropology, particularly folklore, to argue for the vitality of black culture in its various forms within American modernity. Zora Neale Hurston—a pupil of Boas—echoed anthropologists' prevailing view of blacks as "in the way peoples." She found that "the Negro is not living his lore to the extent of the Indian. He is not on a reservation being kept pure. His negroness is being rubbed off by close contact with white culture."[25] Hurston saw "negroness" as an ever-shifting condition that continually reshaped its social and physiological contours. Conversely, NRC social scientists used physical anthropology to delineate blacks' fitness for industrial modernity. Baker notes that, throughout the 1920s, anthropology remained a segregated discipline, rendering a "paramount practical service to the nation" by helping to establish legal and de facto Jim Crow throughout the nation.[26] Committee members identified five key areas in urgent need of attention: a bibliography of Negro anthropology and psychology; research on development of the Negro child; "investigations of the full blood Negro American and Negro mentality; studies of the Negro in Africa to determine the persistence of racial traits"; and "the collection of families of full bloods and browns for future investigations"—all of which were informed by an eugenic obsession with racial purity.[27]

Postwar inquiries into the nature and prevalence of hybridity and its long-term effects on the body politic was only the latest reconstitution of America's ever-shifting black-white racial binary. The figure of the mulatto came to represent a barometer of racial health, or lack thereof. The war and the immediate postwar era saw race and racial fitness increasingly connected to color and the body. Yet postwar anthropology represented a slight detour in this process at a time when strict Mendelian models of biological determinism dominated debates regarding race and racial difference. NRC anthropologists often unwittingly embraced a form of neo-Lamarckism regarding the malleability of racial types through acquired traits. Convinced that "all the nation's economic, social, and political problems were intimately bound up with the reactions of different peoples in our midst," anthropologists conceded that said reactions were born of a social context rather than any innate traits.[28] These "reactions" and the resulting instances of hybridity pointed to the fact that perhaps race was not a fixed phenomenon. Yet committed nativists and eugenic advocates like Madison Grant were unmoved, warning lawmakers, "It has taken us fifty years to learn that speaking English and wearing good clothes and going to school does not transform a Negro into a white man" and no amount of social engineering could obscure this fact. Grant and his ilk remained suspicious of the prewar 'near-white races' currently navigating the fractious postwar road to whiteness. Just like the Negro, "Americans will have a similar experience with the Polish Jew, whose dwarf stature, peculiar mentality and ruthless self interest are being engrafted upon the stock of the nation."[29] Distinguishing between "colored work" and "white man's work" was essential for the long-term health of the nation's labor economy. NRC officials hoped to use the results gleaned from wartime testing to build the case that national efficiency was contingent upon the regulation of shifting cultural and biological racial demographics through legislative means: fixing the right races to the right occupational niches.

The NRC and the Production of Racial Labor Knowledge in Postwar America

The production of knowledge on race and labor was an imperative of the postwar state. The state's facilitation of racial labor expertise functioned as an agent of state power through both institutional and legislative means. From an institutional perspective, these processes of state making through race making began with the founding of the National Research Council (NRC) in December 1916. Eight months prior, President Wilson had charged the National Academy of Sciences with organizing "the scientific agencies of the United States." As the likelihood of America's entry into the war increased, so did the need for a more efficient coordination of the nation's intellectual expertise. In May 1918, the President issued an executive order requesting the Academy to perpetuate the NRC, defining its duties as follows: "To stimulate research and apply the knowledge to the public welfare; to plan comprehensive researches and to minimize duplication by cooperation; and to gather up and render available existing scientific and technical knowledge." The biologist John C. Merriam—who Wilson chose to head the NRC—proposed a science "defined by 'method,' oriented to 'control,' and sustained by organized professional structures to promote research."[30] Methodologies of social control coalesced in the production of a new Negro type that was both a lens of analysis and an object of inquiry through which federally sanctioned "experts" within and outside of the NRC strove to make sense of the shifting demographics of the postwar labor economy.

State-sanctioned social scientific expertise was a public and private enterprise before and after the war. Before the war, the major private philanthropic foundation in the field, Russell Sage, had limited itself to social welfare studies rather than the promotion of the social sciences per se. Jerry Gershenhorn argues that the war reinforced social scientists' move away from a "moral fervor for reform" to a "reverence for scientific knowledge and technological innovation."[31] Following the war, both the Carnegie and Rockefeller Foundations contributed vast sums of money to the development of social science expertise in a variety of public and private mediums. The Department of Research and Investigations at the National Urban League (NUL)—responsible for most of the research and publications on black workers in the 1920s—was initiated by an $8,000 three-year grant from the Carnegie Corporation. Likewise, the Social Science Research Council (SSRC), a public/private initiative, relied on Rockefeller money for its extensive array of summer conferences, advisory councils, and fellowship monies for the study of the social and racial dynamics of postwar political economies. Dorothy Ross argues that this corporate largesse was motivated in part by a faith in the "emerging scientific idiom of the social sciences," along with more cynically minded attempts to purge the family name of prior accusations of corruption, malfeasance, and social injustice. In contrast to the "amateurish philanthropy" of the past, this new idiom promised "both distance from political controversy, and knowledge that would allow the real control of social change"—a decidedly conservative model of social change which would come to inform the public sector.[32]

Conservative models of social change were invariably predicated upon conservative methodologies such as eugenics, which intersected with anthropology through the discipline of physical anthropology. Ales Hrdlicka was largely responsible for making physical anthropology a well-defined field within the discipline through the establishment of the *American Journal of Physical Anthropology* (in 1918) and the American Association of Physical Anthropologists (in 1929). He was also instrumental in securing positions for Madison Grant and Charles Davenport on the NRC's Anthropology Committee. These elite connections provided Hrdlicka with vital political influence. In a 1921 speech at American University, he explained the links between anthropology and eugenics and their implications for public policy: "From now on, evolution will no longer be left entirely to nature, but will be assisted . . . and even regulated by man himself . . . This particular line of activity is known today under the name Eugenics. . . . (which) is merely applied anthropological and medical science—applied for the benefit of mankind." By the early 1920s, social scientists—specifically anthropologists—increasingly turned to migration as a forum in which the laws of biology could be transformed into effective social policy.[33]

Roots and Routes: Anthropology and the Racial Imperatives of Postwar Migration

The sudden appearance of masses of nonwhite workers in the wartime/postwar West engendered a renewed social scientific interest in migration. Migratory patterns were increasingly seen in pathological terms as evinced in observers' over-reliance on terms such as "peril" and "contagion." Stoddard warned that "colored migration is now a universal peril, menacing every part of the white world. Nowhere can the white man endure colored competition."[34] The failure of races to stay in their proper place(s) raised the specter of degenerate germ plasma that infected the body politic of a wider white Western civilization. Yet fears over foreign immigration and the march of the colored races abroad were tied to anxieties regarding the migration of black Americans domestically. President Warren G. Harding remarked, "Whoever will take the time to read and ponder Mr. Stoddard's book on *The Rising Tide of Color* . . . Must realize that our race problem here in the United States is only a phase of the race issue the whole world confronts. Surely we gain nothing by blinking at the facts."[35] Progressives feared that southern black migrants, untethered from traditional social restraints, would fall prey to the menacing specter of "Bolshevism . . . which sought to enlist the colored races in its grand assault on civilization." For "prophets of pessimism" like Stoddard, Bolshevism's insidious and rapacious nature was a product of its "Asiatic mind," which saw the "very existence of superior biological values as a crime." "Bolshevists," claimed Stoddard, "are mostly born and not made." The "great Negro quarters of New York, Chicago, and other northern cities" were seething with ideas and emotions, "which by the power of mass contagion may engender sudden and startling developments." President Wilson had previously worried that black soldiers would be "our greatest medium in conveying Bolshevism to America." White elites correlated fears of labor radicalism and racial pollution, conjuring up nightmares of "red scares, yellow perils, brown hordes, and black problems," which would consume a fragile postwar white supremacy.[36]

Black radicals readily embraced a eugenically informed anti-imperialism in which the "darker masses would eventually overthrow their degenerate white masters." Davarian Baldwin argues that pan-Africanists—such as Hubert Harrison, Cyril Briggs, and Marcus Garvey—consistently denounced the interconnections between "colonialism, global capitalism, racial science, and racist social formations in transnational metropolises." Drafting the manifesto for the 1921 Pan-African Congress, DuBois acknowledged these global processes and his hope for freedom from "the industrial machine and the need to judge men as men and not as material and labor."[37] The era's rapidly shifting labor demographics drove the private lives of a "dark" proletariat onto the public streets of cities throughout the Western world— exacerbating the "problematization" of colored peoples and labor in the postwar social scientific imagination. Tensions over access to services and infrastructure strained urban race relations across the nation. These tensions culminated in a series of bloody race riots during the "red summer" of 1919. Amid the summer's brutal paroxysms of anti-black violence from Chicago to Washington, D.C. to Omaha, white elites generally refrained from viewing these riots in socioeconomic terms and instead cited them as further evidence of the Negro's inherent incapacity for urban industrial modernity.[38]

The NRC established a Committee on Race Characters (CRC) in the spring of 1921 to investigate the racial dimensions of migration. Though short lived, the CRC conducted some intriguing work delineating "the relation of anthropology to Americanization." At the University of Minnesota, Dr. Albert Jenks, Professor of Anthropology, developed an "Americanization Training Course," under CRC auspices, to investigate "the anthropological dimensions of assimilation." He argued that "it was not until America was rudely awakened by a time of national peril that she realized the magnitude of the task before her of assimilating the various people in her midst." Jenks believed, in this respect, "Anthropology has an opportunity for paramount practical service to the nation." Trained to "know peoples, in physiological and psychological terms," graduates would work with social workers, police, and industry to ease the nation's disparate populace into their proper occupational niches. Jenks was especially interested in anthropological research on the Negro, "the least authentically and commonly understood race group in America." As the nation's oldest racial minority, black Americans were seen as an ideal test case for determining the potential physiological and mental effects of assimilation on present-day foreign immigrant groups and the biological dimensions of citizenship.[39]

The CRC's failure to secure proper funding soon led to calls for a new committee. In September 1922, the NRC's Committee on Anthropology and Psychology recommended the establishment of a committee for the scientific study of human migration under the direction of Robert Yerkes, who had directed the infamous Army IQ tests. Two months later, the Committee on Scientific Problems of Immigration (CSPI) held its first conference in Washington, D.C. Some twenty individuals representing a wide range of disciplines from biology, economics, psychology, and anthropology attended the conference. Funded by the Russell Sage Fund and the Laura Spelman Rockefeller Memorial, the committee's goal was to consider, "from the view of natural science, the complex migration situation resulting from the world war and from the virtual elimination of space as a barrier to movements of men and race intermixture."[40] Anthropological evaluations of migrant bodies sought to delineate the racial dimensions of migration as a precursor to the management of these processes.

The CSPI ultimately proved too narrow for the transnational demands of its work. In early 1923, in order to "see the world-situation clearly and without individual, national or racial bias," members dropped their focus on "immigration" and adopted the broader designation, "Committee on the Scientific Problems of Human Migration" (CSPHM). Throughout the spring and summer of 1923, the committee sought to "speedily bring all the resources of science to bear on the study of migration." Members noted that considering the "important and imminent modifications of national immigration policy," knowledge of "human traits and potentialities—individual, occupational and ethnic" was vital. The committee requested a sum of $60,000 from the NRC for studies into the psychological, anthropological, and socioeconomic factors of migration. The latter two connected matters of "racial physical characteristics, normal and pathological" to "immigration's relation to labor supply and its distribution and relative adequacy among the different industries, trades and arts." CSPHM members became convinced that "one could not form a judgment on the problem of immigration so far as it concerns the United States without a consideration of the race problem in the broadest sense." Members were committed to undertaking "a comprehensive biological evaluation of the processes going on in America resulting from the inflow and subsequent assimilation of a wide variety of racial groups"—revealing the ongoing connections between race and nation in U.S. immigration policies.[41]

Racial "roots" and migratory "routes" were linked through the emerging discipline of climatology, which was broadly defined as the study of the effects of climate on racial development. The pioneer of modern climatology was Yale geographer Ellsworth Huntington. CSPHM officials at the NRC-funded Committee on the Atmosphere and Man drew on Huntington's work to produce a racial geography of civilization from the tropics to the tundra.[42] Prewar studies of race and climate—such as Frederick Hoffman's analysis of blacks in Canada—deeply informed these new environmental models of racial difference. For instance, Hoffman found that in 1901, the black population of Canada was 17,437 versus 16,877 in 1911. In that time, the aggregate population of Canada had increased "and the environmental conditions were probably never as favorable to Negro progress in the Canadian Dominion than they are at the present time." Hoffman concluded that "the race is not holding its own; it is not increasing by an excess of births over deaths and in Canada as in the U.S., is distinctly subject to an excessive disease liability and mortality due to its inherited race traits." Negroes' experience in Canada proves conclusively that there is no tendency on the part of the Negro to migrate to far northern latitudes nor any inherent power of successful adaption and race survival."[43] Given the logic of environmental determinism, races out of place were invariably races in decline.

NRC social scientists saw race—born of specific environmental and climatic conditions—as the driving force of migration. The instinct to wander, or nomadism, was one with a hereditary basis. While previous immigrants from northern Europe had made a conscious decision to emigrate, the present day masses of colored laborers were seemingly driven by an almost primal urge to seek new opportunities. While the former represented the hardiest stock of the Nordic race, the latter were the "dregs of humanity."[44] Francis Amasa Walker was adamant that modern migratory patterns were a function of failure: "They are beaten races, representing the worst failures in the struggle for existence."[45] Dan Bender argues that "climate, race, and migration transcended the nation and permitted scholars to rationalize the racial kinship they felt with their European counterparts," in opposition to the colored hordes. Climatology crossed national boundaries in linking together a common white race.[46] Developing a racially informed immigration policy required an intricate catalogue of racial taxonomies for the "wise regulation of mass movements of mankind" and the "safe development of social biology" at home and abroad.[47] By 1919, uncertain socioeconomic conditions, labor unrest, and a rising eugenic mindset had led to a broad public and social scientific consensus in favor of immigrant restriction—though the form, degree, and duration of such a policy remained in dispute.

NRC members figured prominently in the deliberations of the House Committee on Immigration and Naturalization, chaired by Congressman Albert Johnson of Washington. Johnson was a former AEF officer, a virulent anti-Communist, and the president of the Eugenics Research Association (ERA). Under Johnson's leadership, the ERA had vigorously lobbied state and federal authorities for immigration restrictions on "inferior" racial stocks, forcible sterilization of the disabled, and a permanent ban on interracial marriage. Johnson hired Harry Laughlin of the Eugenics Records Office—and disciple of NRC member Charles Davenport—to ensure the congressional deliberations maintained a racial focus. As the committee's expert on race biology, Laughlin testified, "The character of a nation is determined primarily by its racial qualities; that is by the hereditary, physical, mental, and moral temperament of its people." Throughout his testimony, Laughlin was also in close correspondence with Yerkes's student, Professor Carl Bingham, who was currently conducting a study into the "internationalizing or universalizing of methods of mental measurements" under the auspices of the CSPHM. Like Yerkes, Bingham took a positivist view of intelligence testing, believing it to be a latent, definable racial trait—a definitive index of racial fitness for American citizenship. Citing the results of wartime intelligence testing, Laughlin argued that the melting pot could not be "allowed to boil without control" in blind pursuit of the national motto that declared all men to be equal.[48] Laughlin impressed upon Congress that the health of the republican body politic depended on delineating which peoples could be afforded equality—and the exclusion of those who could not—through strict racial quotas rooted in hereditarian models of racial development.

The Johnson-Reed Act of 1924 represented a profound shift in national immigration policy from a period of relatively open European immigration to one of greater restrictions on the former and the reiteration of Asian exclusion. Mae Ngai notes that the Act "articulated a new kind of thinking, in which the cultural nationalism of the late nineteenth century had transformed into a nationalism based on race" with the broadly defined white Caucasian at its apogee. Perhaps the most significant change was in the introduction of a quota system, which although designed to restrict immigration from southern, central, and eastern Europe, "also divided Europe from the non-European world. . . . It defined the world formally in terms of country and nationality but also in terms of race."[49] Davenport remarked that "before the war it was generally regarded as impractical to make a selection of the fit on the other side. The war has now shown the possibility of doing so." Yet even the bill's most ardent advocates such as Lothrop Stoddard did not see it as a definitive solution to the immigrant invasion, but as "the beginning of a new epoch of national reconstruction and racial stabilization that would culminate in a *reforged America*."[50] Managerial elites agreed that, given the apparent unsuitability of nonwhites as long term sources of labor, the nation should concern itself with "making that which we have doubly efficient." The multitude of near-white races of the prewar era could now be transformed into sturdy and responsible Caucasians to stem the rapid flow of colored labor.[51] DuBois, in an effort to forge a cross-racial international proletariat, urged black workers that "we will make America pay for her injustice to us and to the poor foreigner by pouring into the open doors of mine and factory in increasing numbers," linking African Americans' rupture of the Mason-Dixon Line via mass migration to the closing of the doors of Ellis Island to so-called immigrant undesirables.[52]

Wartime testing of blacks was ambiguous regarding their labor and military fitness. The institution of draconian "work or fight" laws, aimed primarily at poor southern blacks, reinforced the notion that the Negro would not work free of coercion. This anxiety grew as southern black migrants moved north into wartime industries that lacked the traditional constraints of Jim Crow. National Urban League officials noted that while many employers conceded the "adequacy of Negro labor," others cited "the race's natural inertia," "unreliability," and "shiftlessness for high speed production" to rationalize pushing blacks out of their "temporary" positions following the war. Officials at the War Department confidently declared, "clearly the Negro has failed to adjust himself to industrial life and must be returned to his natural occupational environment in the South." NUL officials conceded, "The readjustment of the Negro to northern conditions is a difficult task, of course, that must be continued and intensified if the position gained during the war is not to be lost."[53] Yet in the face of the postwar consolidation of the white working class, the status of black labor became ever more tenuous.

Notwithstanding the checkered results of wartime testing blacks reintegration into the postwar economy was further imperiled by anecdotal assessments of their lack of martial vigor. Despite official pronouncements by General John Pershing that blacks had "measured up to every expectation of the Commander in Chief," blacks' disproportionate consignment to labor battalions demonstrated that expectations had been quite low to begin with. Almost immediately following the armistice, black soldiers were subjected to taunts and ridicule in the mainstream press by many of their former white allies who claimed that "the darkies had merely smiled their way through the war." Mainstream papers were filled with comic accounts of black troops fleeing from their first encounter with shelling and expressing befuddlement with foreign French ways, all the while shuffling and shucking in a stereotypical "nigger" dialect.[54] In 1925, Retired General Robert Lee Bullard, commander of the 92nd Negro Division, caused a stir when he stated that along with being "intractably lazy" and "wholly lacking in initiative," the men under his command generally had little idea of what was expected of them as soldiers. Bullard concluded that if one "needed combat soldiers, and especially if you needed them in a hurry, don't put your time upon Negroes."[55] Black veterans were consistently maligned for their seemingly congenital lack of willpower. Much like their counterparts in the industrial sphere, military officials argued that black soldiers required harsh military discipline to compel them to fight. Army officials justified the need for segregated units in part because of blacks' supposed lack of soldiering qualities and because they feared that the race's "natural torpor" would infect the ranks if left unchecked. The trope of "Negro lassitude" was cleverly inverted by the editors of the *Eugenical News*, who claimed, "The worst thing that ever happened to the area of the present United States was the bringing of large numbers of Negroes, the lowest of races, to our shores. America called for cheap labor that its whites might enjoy the luxury of the parasite, which is fed by its host without effort of its own. Now we realize that this host bids fair to destroy the parasite."[56]

White Americans' parasitic reliance upon Negro labor had now seemingly left the former vulnerable to the degradation of the latter.

Conversely blacks—especially those schooled in the Tuskegee/Hampton model of vocational uplift—challenged this thinking by casting black labor as a bulwark against the potential deluge of foreign labor poised to stream out of war-ravaged Europe. While Stoddard cited "the profoundly destructive effects of colored competition upon white standards of labor and living," Kelly Miller characterized black workers as able-bodied and law-abiding American citizens who required no "Americanization." Drawing on the prewar rhetoric of Washingtonian uplift, Miller noted that it was unnecessary to look to foreign shores to offset any labor shortage when such a "large and sympathetic group was within reach."[57] When officials at U.S. Steel released a statement deploring labor shortages due to present and pending immigration laws, Tuskegee's Emmett Scott proposed the use of black workers. In contrast to "defective foreign labor," the Negro worker was "not an alien, he possessed a strong body and a real attachment to American institutions." The NUL lamented that employers "were ill prepared to make use of Negro labor," instead opting for "white labor of all ranks, even the most unskilled and ignorant foreigners who greet our arrival with widespread and unreasoning hostility."[58] Antiblack racism on the part of immigrant laborers revealed the imperatives of whiteness—both as ideology and identity—in the formation of the postwar labor economy. The prospective wages of whiteness outweighed any potential benefits of cross-racial worker solidarity.

"City Negroes": Urbanization and the New Negro Type in Postwar America

Tracing the various migratory routes of peoples in the wartime and postwar era led to a concurrent interest in the effects that urban destinations had on racial and labor identities. The postwar New Negro type was an urban creation. The 1920 Census revealed that, for the first time, more Americans lived in cities than in rural areas. Between America's entry into the war and the stock market crash of 1929, blacks left the South at an average rate of five hundred per day, or more than fifteen thousand per month. By 1930, more than a million blacks had left the region of their birth.[59] The era's sprawling metropolises were draining the world's vast open spaces of capital and peoples. Through a "strain of peculiar racial status and the terrific pressure of modern life," the city created its own types—perhaps the most prominent of which was the "Negro" who, in little over a decade, would come to be a synonym for "urban."[60] Robert Park claimed that black migration out of the "caste-ridden rural South" and into cities of the North and Midwest meant that their situation could no longer be depicted as a "natural history" but would now have to be investigated as "social history." Whereas the rural environment required an organically informed analysis, an examination of the complexities of urban centers necessitated social scientific methodologies. Postwar urban modernity—through shifting labor processes and a rapid expansion of consumer culture—fundamentally altered the temporal and spatial dynamics of urban dwellers across the color line.[61]

Longstanding notions of the city as a site of social and racial degeneracy continued to inform urban theory in the postwar era. The dysgenic effects of war had provided elites with the language of eugenics, which linked an earlier rhetoric of urban reform that located degeneracy in social causes to one rooted almost exclusively in heredity. For Madison Grant, the so-called Nordic type was handicapped by a natural aversion to industrialism and urban life: "The cramped factory and crowded city quickly weed him [the Nordic] out, while the little brunet Mediterranean can work a spindle, set type, sell ribbons or push a clerk's pen far better." According to Grant, the "somewhat heavy Nordic blond who needs exercise meat and air cannot live under Ghetto conditions."[62] Shortly following the armistice, Warren Thompson of Cornell University noted, "City life seems to be unfavorable to the raising of even moderate sized old stock families among all except the poor." City life diminished the vitality of the better classes and races while exacerbating the proliferation of the lower peoples and deviant types such as blacks.[63]

Nor was the black intelligentsia immune to eugenic interpretations of urbanization. Postwar black observers drew heavily on prewar investigations into race and urbanism such as the Atlanta University Studies of the 1890s. Led by R. R. Wright, the Atlanta studies were an early effort by black social scientists to investigate potentially causal relationships between biology, urban poverty, and vice. Though environmentalism eventually carried the day over genetics, researchers at Atlanta failed to achieve complete consensus. Those that rejected environmental causes tended to use class and cultural analyses, citing working class blacks' failure to measure up to white middle class norms of respectability. Eugene Harris of Fisk University, in "The Physical Condition of the Race: Whether Dependent Upon Social Conditions or Environment," posited that a genetic disposition rooted in slavery and manifested in class as the reason for blacks' failure to acclimatize to modern urban life. Even DuBois's *Philadelphia Negro* cited urban living as a contributing factor in the proliferation of the degenerate traits and tendencies of "the lower classes of Negroes" in cities, effectively reconfiguring degeneracy as a function of class rather than race.[64]

Postwar progressives viewed the socioeconomic and racial dislocations of migration and urbanization in deeply historical terms. Yet historian Joe W. Trotter Jr. notes that while scholarship during the Great Migration "placed black migration within a larger historical context," this often came at the "expense of examining this process in depth over time." Blacks' migration north was seen as only the most recent example of the migratory impulse that had characterized the nation's history, but the particulars of the Great Migration were generally given short shrift. Postwar anthropologists tried to reconcile this disconnect between the past and the present by charting black migration and urbanization in physiological or corporeal terms. The editors of *The Survey* characterized the migration and urbanization of "tropical colored laborers" out of their natural rural southern habitat as a "physiologically violent act" with irrevocable historical consequences for future race development.[65] For many observers, blacks' encounters with the stimuli of the modern city revealed themselves in explicitly physiological terms. Anthropologists such as Melville Herskovits unsuccessfully attempted to measure a cross section of Harlem's inhabitants "to assess the short and long term effects of migration on Negro Physiognomy." Building on Boas's earlier work on changes in immigrant physiognomy, Herskovits anticipated that environment, not biology, was the determining factor in blacks' urban development. Herskovits's methodological impetus revealed a shift among anthropologists, from hereditary to cultural models of racial difference and development in urban contexts.[66]

During the war, many black social scientists began to reject the links between urbanism and racial degeneracy—characterizing the "increasing urbanization of the colored race as an inevitable process" and proof of racial progress rather than decline. Led by Charles Johnson, the NUL's Department of Research and Investigations, along with *Opportunity* magazine, were at the forefront of research on postwar black workers. Founded in January 1923, *Opportunity* was home to a new generation of black artists, social scientists, and social workers. Along with providing a forum for fledgling literary talent, the magazine represented a break from prewar reformism through its "desire to approach . . . new problems with a new increased scientific technique for dealing with them," which essentially reified the Negro as object—albeit through the lens of black folks—of social scientific inquiry.[67] Efforts to come to terms with blacks' on-going marginalization from mainstream urban labor economies led many black social scientists to cite both heredetarian and cultural causal factors. Charles Johnson in *The Black Worker in the City*, argued that blacks, "by tradition, and probably by temperament," were the antithesis of the modern urban type as their "métier was agriculture." The laboring black body was literally incompatible with modern work processes, given "the in-complex gestures of unskilled manual labor and even domestic service; the broad, dully sensitive touch of body and hands trained to groom and nurse the soil, develop distinctive physical habits and a musculature appropriate to simple processes." Moreover, Johnson saw these processes in explicitly spatial terms, claiming, "It is a motley group which is now in ascendancy in the city. The picturesqueness of the South, the memory of pain, the warped unsophistication, are laid upon the surface of the city in a curious pattern." Conceptions of blacks' incapacity for urban life were a sharp rejoinder to prewar and wartime models of the Negro as benefitting from the transition to urban life advocated by the likes of George Haynes. Alain Locke posited this cultural model of black migration as "the Negro's deliberate flight not only from the countryside to city, but from medieval America to modern"—which had seemingly altered the physiological and psychological contours of blackness in the process.[68]

Postwar migration and rapid urbanization also reopened prior debates regarding black extinction. Black birth rates—a key metric in Hoffman's thesis of blacks' imminent extinction—once again became a topic of public discussion. Following the war, however, many believed that popular ideas "about Negro health and extinction had suffered severe shock in the light of improved medical science." According to the U.S. Census Bureau, the number of blacks nearly doubled between 1870 and 1920, leading the editors of the *World's Work* to confidently declare, "Death now awaits the hope that the Negro will die."[69] Between 1900 and 1910, the decennial increase in black population was around 11%, with ten million black Americans making up roughly 10% of the nation's population in 1910. Although this rate was commensurate with whites—both native and foreign-born—the black birth rate still remained well below that of whites for the next ten years. Whether due to birth control, poverty, war, clerical error, or the postwar influenza epidemic, the 1920 Census revealed that African Americans entered the third decade of the century at just under 10% of the national total. Notwithstanding blacks avoidance of extinction, the race appeared to have reached a point of stagnation.[70]

To reinvigorate the health of the race, black reformers embarked upon a number of key initiatives. Foremost of which was the NUL's "National Negro Health Week," launched in conjunction with the U.S. Public Health Service. Negro Health Weeks joined agencies such as the National Negro Business League, the American Social Hygiene Association, American Red Cross, and the YMCA to promote health as a basis for occupational and economic advancement. Health was the key to not only "reduce the cost of preventable disease and death, but to also increase vitality, resistance to disease, and the well being, earning capacity, and service of the healthy citizen to home, community, and country." Held annually during the first week in April—to commemorate the birthday (April 5) of Booker T. Washington, officials from these various agencies held workshops, lectures, and film screenings to educate northern black communities throughout the nation on the importance of hygiene and sanitation at home and in the workplace. Prior moralistic approaches to health were slowly being displaced by ones which saw health in mechanistic terms and "the human being as living machine," which required the need to place health and hygiene along "modern scientific lines" to promote national efficiency.[71] Developing health capital was instrumental in the NUL's broader strategy of promoting efficiency as the key to eliminating workplace racism. A fitter worker was a successful worker with greater occupational opportunities. NUL officials argued that the limited gains made by black workers during the war demonstrated the ability of market forces to mitigate prejudice—thereby making the systematic distribution of health expertise and vocational training essential to racial uplift.[72]

Paradoxically, postwar theories of black extinction were most prevalent within black communities. As late as 1924, *The Forum* contended that black migrants to the North would "gradually die out, for there he seems to lose his fecundity." Marcus Garvey concurred, claiming that "the Negro is dying out, and he is going to die faster in the next fifty years." The historian Carter Woodson conceded that on "account of this sudden change of the Negroes from one climate to another and the hardships of more unrelenting toil, many of them have been unable to resist pneumonia, bronchitis and tuberculosis." However, Woodson was quick to point out that many of the reports on black migrants' poor health had been greatly exaggerated.[73] The mainstream press peppered analysis of the Great Migration with accounts of migrants freezing to death on the cold and unforgiving streets of Northern cities. NUL officials agreed: "It is a strange fact that in the cities of the North, the native born Negro population, as if in biological revolt against its environment, barely perpetuates itself. For whatever reason, there is lacking that lusty vigor of increase which has nearly trebled the Negro population as a whole." Within the past sixty years, the natural increase of this old Northern stock, apart from migrations, has been negligible.[74] Black nationalists characterized this alarming lack of fecundity as a function of economic pressures and a lack of racial purity. The editors of the Universal Negro Improvement Association's *Negro World* claimed that the remedy for blacks' inferior socioeconomic position called for "clean and orderly sexual relations" as a basis for a self-sufficient black capitalism.[75]

Throughout the immediate postwar era, UNIA members infused the discourse of contemporary political economy with the tenants of a eugenically minded black nationalism—creating a narrative of racial uplift predicated on the inviolability of black bodies. In contrast, UNIA critics such as DuBois and Kelly Miller couched their eugenic impulses in the intra-racial rhetoric of class—inveighing against the dangers of "the masses" out-breeding "the classes." Mia Bay notes, "Ironically, just as the white scientific establishment was finally beginning to dismantle the racial edifice built by nineteenth-century science, racial essentialism was achieving unprecedented popularity in some corners of the black community," such as the UNIA and various other nationalist organizations.[76] Linking economic uplift with racial integrity, Garvey warned that "if we do not seriously reorganize ourselves as a people . . . our days in civilization are numbered." Garvey's sentiments revealed a vision of a martial and masculine "race never conquered" produced and sustained by an undiluted blackness.[77]

The New Negro, the Caucasian, and the Mulatto, and the Rise of Postwar Biracialism

Postwar debates regarding the effects of migration and urbanization on black health presumed a clearly defined Negro "type." This taxonomic mindset invariably raised questions regarding the physiological contours of blackness and the perils of hybridity. Anthropologists were especially interested in tracing racial types through the various stages of industrial evolution. For A. E. Jenks, the "question of so-called disharmony in crosses demands attention more urgently than almost any other, as no nation can survive except under the conditions nature imposes."[78] Rising racial and labor unrest at home, anti-colonialism and bolshevism abroad, along with a prevailing biological mindset transformed local discourses of "miscegenation," or race mixing, into a transnational phenomenon. Recent scholarship on American empire has elucidated the links between domestic anti-miscegenation statues, their colonial counterparts, and the respective role of each as a transnational node of racial labor control. At home and abroad, these methods of sexual surveillance were couched in the rhetoric of hygiene and carried out under the auspices of public health programs.[79] In the fall of 1926, the NRC began arranging a Committee on the American Negro to focus on the "anthropological and psychological dimensions of the Negro." The real attention, however, was on "how far race mixture and race contact may affect social hygiene and our civilization." Though blacks made up approximately one-tenth of the population, a supposed rise in race mixing had led to blacks being "perhaps the least known of any part" of the nation's peoples.[80] NRC officials cautioned that attempts to "forge a nation out of disparate peoples constituted a ceaseless warfare with the stubborn biological forces of nature." Victory in this conflict would only be achieved through the preservation of racial integrity and purging any and all racial transgressions from the republican body politic.[81]

The Committee on the American Negro was a "who's who" of the era's self-described race experts from Charles Davenport, Ales Hrdlicka, E. A. Hooton, to committee chairman R. J. Terry. Even Franz Boas—the leading purveyor of anti-racist social anthropology and scourge of the eugenics set— actively sought out membership and was present at the committee's founding meeting in Philadelphia in December of 1926, where he seconded Davenport and Hrdlicka's recommendation that the committee create a "bibliography of prior and current research on the American Negro."[82] Though the committee retained a commitment to original research on race mixing, many felt proposed historiographical or bibliographical commitments of this nature were beyond their financial and institutional means. Instead, members committed themselves to the consolidation of knowledge on "the biological, psychological, and physiological aptitudes and aspects of the full-blood Negro and of the mixed population," which had proven resistant to the imperatives of military and industrial systems of racial classification.[83]

Throughout the Progressive Era, the cultural practices of endogamy— marrying and procreating within one's specific class, caste and or ethnicity— were rigidly observed and enforced across the color line to preserve the social peace. Negrophobes and black nationalists found common cause in the belief that race crossing was socially and biologically detrimental to a race's progress. In the fall of 1923, Marcus Garvey wrote to John J. Davis, the Secretary of Labor, on the subject of the "Negro problem." Noting that "the twentieth-century Negro in America was different from the Negro of the last century," Garvey queried Davis on the department's potential commitment to blacks' repatriation to Africa. Linking race purity to industrial advancement, Garvey asked Davis whether he believed "that the Negro should be encouraged to develop a society of his own for exclusive social intercourse" or to "create positions of his own in industry and commerce in a country of his own." Though departmental officials did not pursue the matter further, the correspondence reveals the power of eugenic interpretations of labor economy. Historian Michelle Mitchell notes that postwar "uplift ideology adapted quite readily to the period's characteristic eugenic thinking," specifically in Garveyite calls for racial purity as a means to a race's socioeconomic and cultural autonomy.[84]

Recent historical studies have revealed a concurrent—though decidedly minority—strain of thought which saw race crossing as a positive practice, which could restore national vitality.[85] Motivated by both personal and political considerations, DuBois had long railed against characterizations of the enfeebled and infertile mulatto. In an era that equated racial vigor with manliness, however, this proved a difficult discursive task. Legislative efforts to strengthen the black-white binary turned the mulatto into a legal impossibility and a social pariah. Franz Boas, in his anthropological work for the prewar Immigration Commission, had argued that race mixing was the only way to instill a much-needed vitality in "an industrially and socially [not biologically] inferior black population." Years later, in a piece entitled "The Problem of the American Negro," Boas reaffirmed his belief that since racism rested on social awareness of differences—exacerbated by economic competition—then the solution was to diminish these differences as much as possible. But even Boas, the famed anti-racist and cultural relativist, saw amalgamation as a one way street which ultimately entailed the disappearance of black Americans into a colorless, and by default white, mass populace. The proposition that race mixing could "blacken" the populace in a positive manner was simply beyond the pale for the majority of social scientists. For Boas and many of his progressive peers, the mulatto was merely a detour on blacks' inevitable road to social and perhaps even physiological whiteness: race would cease to be a problem with the elimination of blackness in all its forms.[86]

Foregrounding the mulatto as an object of inquiry revealed prevailing anxieties over the stability of racial categories, revealing key cracks in the leviathan of postwar white supremacy. In 1890, the Census Bureau made its first and only attempt to divide peoples of African ancestry along the lines of "negro," "mulatto," "octoroon," and "quadroon." Following 1920, the category of "mulatto" disappeared altogether from the federal census. The editors of *Opportunity* ruefully remarked, "Men who, by and by, ask for the Negro will be told—'there they go, clad in white man's skins.'"[87] Anthropologists dismissed arguments for the mulatto as a new or distinct race, arguing that mixed race individuals "possessed a low degree of physical variability" and would eventually be subsumed into the larger mass of African Americans. Speaking to the NRC Conference on Racial Differences, Dr. A. Cole of Columbia remarked, "If we do find changes occurring in bodies of individuals who make up our immigrant groups; if our Negro is changing; if blending is going on, it must lead to radical readjustments in our concept of race."[88] Defining the mulatto as Negro—using the one-drop rule of blackness—required shifting the physiological and social boundaries of both blackness and whiteness.

Wartime assessments of decommissioned soldiers only served to muddy the color line. Testing revealed that mulattos possessed a more "well defined musculature and proportionality of appendages" than whites. Moreover, all blacks—both mixed and "pure bred"—had many physical advantages over whites. Blacks, broadly defined, were much less prone than whites to suffer from defects of the spine, obesity, deaf mutism, deafness, and diseases of the eyes, nose, and throat. Mulattos' disproportionately high scores on the Army IQ tests recalled previous theories of racial hybrids' mind and body imbalance: the superior white intellect captive to the savage black body. Davenport characterized mulattos as more "restless on the whole than the Negro and less easily satisfied with his lot—possibly due to a disharmony introduced by the cross."[89] A restlessness that was seemingly symptomatic of white's inborn independence of spirit, which was continually stifled by the crushing "lassitude of the mulattos" Negro blood. Robert Park diagnosed the mulatto as a "marginal man" consigned to an ineffectual existence because of his inability to join the supposedly "superior white group" or create a separate caste from the "pure-blooded Negro."[90] Historian Greg Carter notes this burgeoning school of "mulatto studies" analyzed hybridity to assess the progress of their minority communities—that is, blacks—as opposed to the progress of America as a whole: the mulatto as a barometer of blacks' racial health.[91]

Models of the "maladjusted" mulatto sought to delineate racial difference in physiological and psychological terms. NRC officials were especially interested in charting the race's respective mental "aptitude for modern civilization."[92] In the mid-1920s, Dr. Joseph Patterson of George Peabody College in Nashville, Tennessee, under the auspices of the Committee on Human Migration and Race Characters, conducted studies on local school children to gauge the relation of intelligence to race difference, or what he deemed "racial ingenuity." Patterson ignored Yerkes's wartime caveat that "these intelligence tests do not measure occupational fitness nor educational attainment; they measure intellectual ability which has been shown to be important in estimating military value."[93] Notwithstanding mulattos' high test scores on the Army IQ tests, Patterson believed that intelligence testing engendered "a naturalistic attitude in a community toward behavior and the success or failure of individuals and or groups in particular." As people came to feel that group accomplishment was based "upon innate characters with some degree of training, rather than the haphazard factors and arbitrary volitions of individuals," racial hierarchies could be preserved. A committed eugenicist, Patterson felt that interracial and intra-racial testing was key to allowing various races to "regulate the selection of factors for the production of desirable types," thereby eliminating "maladjusted strains" like the mulatto.[94]

Postwar characterizations of the mulatto revealed much about the instability of the color line both at home and abroad. Anthropology as "a science of mourning" was fittingly preoccupied with analysis of the mulatto as a legal, social, and biological anachronism. Anthropologists had been among the first to situate the mulatto and hybridity more generally in a transnational context in works such as Eugene Fischer's *Rehobother Bastards and the Problem of Miscegenation Among Humans*; L. R. Sullivan's *Half Blood Sioux*; and *Distribution of Stature in the United States* by Columbia's Clark Wissler. A compatriot of Boas, Wissler believed racial admixture to be rampant throughout the American population, even going so far to put "the negro" of Davenport's *Army Anthropology* in mocking quotation marks. Subsequent studies in the field included *Mestizos of Kisar* by E. Rodewaldt; Leslie Dunn's *An Anthropometric Study of Hawaiians of Pure and Mixed Blood*; and Davenport and Morris Steggerda's *Race Crossing in Jamaica*.[95]

Transnational NRC investigations of "race crosses" were informed by a desire to balance blacks' newfound—albeit limited—mobility with their traditional social strictures both at home and abroad. Davenport argued that despite the body's "great capacity for self adjustment, it fails to overcome bad hereditary combinations." Mixed races, or "hybridized people," were a "badly put together people, a dissatisfied, restless, ineffective people." The lingering influence of climatology led social scientists to continue to characterize the sluggish pace of labor in the tropics with racial admixture. The deleterious effects of race mixing on a society's industrial evolution were made evident in "the indolent republics and colonies" of the Caribbean and of Central and South America. Latin America offered a good "field for the study of racial crosses . . . as the social environment is much more uniform, there being in some countries little or no race prejudice and discrimination of the sort that is alleged to prevent the Negro and the Mulatto in the U.S. from 'coming into their own.'"[96] Whereas de facto and de jure racial prejudice in the United States functioned as a form of self-preservation on the part of whites, the absence of these same mechanisms throughout the Caribbean and Latin America had produced a depraved mixed population unable to compete in the struggle of industrial evolution.[97]

Postwar anthropologists saw the tropics as a laboratory where the natural laws of racial heredity prevailed, unfettered by the artificial social restraints of modern industrial civilization. The prevailing belief among social scientists held that the "primitive races provided good opportunities for strictly scientific studies of mortality problems. Until such races come into contact with civilized man, they generally present a healthy, robust, and vigorous appearance."[98] In March 1926, the Carnegie Institution of Washington accepted a gift from an anonymous donor who requested an investigation into "the problem of race crossing, with special reference to its significance for the future of any country containing a mixed population." Carnegie's Department of Genetics, in conjunction with the NRC, appointed an advisory committee of Davenport, E. L. Thorndike, and Robert Yerkes's former assistant, W. V. Bingham. Morris Steggerda, a promising young zoology student at the University of Illinois "with excellent training in genetics and psychology" was appointed chief field investigator. In the summer of 1926, Steggerda and Davenport traveled to Jamaica to see firsthand the effects of race mixing on the nation's health.[99]

Jamaica's proximity, common language, varied demographics, and cooperative colonial authorities made it an ideal locale to chart the effects of hybridity on a nation's labor economy. American anthropologists had conducted fieldwork on the island for years. Melville Herskovits had conducted extensive research in Jamaica and was largely responsible for convincing the NRC and the Carnegie Foundation to agree to Davenport and Steggerda's proposal for research on the island. Following his study of Hawaii's unique racial dynamics, the ubiquitous Hoffman had visited Jamaica and Cuba in late 1916 to conduct an investigation into tropical mortality. He found that much like blacks stateside, blacks in the islands suffered from poor respiratory health due to supposedly innate physiological deficiencies like diminished lung capacity. Hoffman also noted that rates of tuberculosis and syphilis in Jamaica were indeed comparable to those among American blacks. Yet he was far more willing to ascribe Jamaican blacks' high rates of tuberculosis to the poor and unhygienic living conditions in which the island's colored majority lived as opposed to any "innate" racial traits that afflicted American blacks stateside. Nevertheless, at Hoffman's urging, Prudential Life Insurance subsequently declined premiums to the non-white populace of the West Indies well into the late 1940s.[100]

The NRC's plans for Jamaica were ambitious, and included "a study of Negroes and mulattos in comparison with whites living in the same locality, with special reference to their innate qualities, fitting them for carrying on modern civilization." Along with anthropometric and mental measurements, Steggerda was eager to obtain samples of basal metabolism—a measurement of oxygen consumption, blood pressure, and body temperature believed to be the best index of an individual and presumed racial level of required caloric intake and respiratory capacity. At the outset, Steggerda suspected that the "racial metabolism of Jamaican blacks and browns" would be much lower than that of whites from the temperate zone, confirming the former's apparent "tropical lassitude." Given that 85% of the island's population lived in rural districts working in the production of sugar and other foodstuffs, measurements of agricultural workers were key for producing a representative population sample. Davenport believed that because "the entire population depends—in one manner or another—upon the fruit of the soil for its live-lihood, this undoubtedly has an effect upon the physical measurements of the population." The bulk of initial measurements were to be conducted at the island's schools and municipal institutions—spaces that "lent themselves most readily to scientific work"—before proceeding to an analysis of the ag-ricultural communities.[101]

Upon arrival in Kingston, Davenport and Steggerda made contact with the local colonial authorities. Officials at the U.S. State Department put them in contact with the island's Assistant Colonial Secretary, Superintending Medi-cal Officer, the Director of Education, and officials of the Jamaica Hookworm Commission of the Rockefeller International Health Board. The Americans toured the Kingston Hospital, penitentiary, and lunatic asylum and were assured that staff members and inmates would both be made available for measurements. Davenport and Steggerda found their keenest supporter in H. J. Newman, principal of Mico Training College. Founded in 1834, Mico College was a small training school for male teachers modeled along the lines of Hampton and Tuskegee, although with less emphasis on vocational train-ing. Steggerda and Davenport measured the entire student body of Mico—"fifteen Negros and forty-six mulattos"—and used the school as the primary site of their operations due to its central location and capacity to house the bulky apparatus required for measuring basal metabolism.[102] Subsequent an-thropometrical work in the surrounding communities was ultimately hin-dered by a lack of staffing. From September 1926 to October 1927, Steggerda conducted his limited fieldwork with occasional help from Davenport, who frequently shuttled between Jamaica and the States. In December 1926, Steg-gerda returned to Cold Spring Harbor and presented his initial findings to the NRC and Carnegie Institute fellows in New York. Following these meetings, it was arranged that physical anthropological data should be collected for fifty adults of each sex of three groups—"pure-blooded negro, mulatto, and white, from the same social or occupational level if possible."[103] Steggerda was also to acquire data for a developmental series exclusively on Negroes and mulat-tos. In January, Steggerda returned to Jamaica, where he spent the next ten months traveling throughout the island measuring all variety of its inhabit-ants. Steggerda's final reports totaled approximately eight thousand sheets and scored as they were received. Codes for each trait were tabulated and adapted to Hollerith punch cards similar to those used in wartime anthropometric evaluations of recruits.[104]

Davenport and Steggerda's findings were published in *Race Crossing in Jamaica*, which focused on five factors: evidence of increased variability in race characters; evidence of dominance or recessiveness; appearance of new qualities; appearance of social traits; and evidence of "hybrid vigor." Summarizing their findings, Steggerda and Davenport found that among the traits in which blacks and whites differed genetically, "Browns" (peoples of mixed race) were quite variable. No evidence of dominance or recessiveness of any particular traits was found, nor did Jamaica's Negros exhibit any new genetic qualities or mutations. Given the seeming persistence of African traits and tendencies, "the burden of proof is placed on those who deny fundamental differences in mental capacity between Gold Coast Negros and Europeans." In terms of mental capacity, Browns on average were found to be "intermediate in proportions between whites and blacks . . . though an excessive percent seemingly failed to be able to utilize their native [read: white] endowment." Assessing racial hybridity, the authors curtly concluded, "no evidence of hybrid vigor is found in Browns," confirming a priori notions of hybridity as a source of racial degeneration.[105]

Anthropological fieldwork in Jamaica reaffirmed prevailing characterizations of hybridity as a degenerate social force, but was occasionally challenged by those who saw the practice as potentially invigorating to a depleted racial stock. Heterosis, or "hybrid vigor," occurred when the union of two races— both of which were unable to express their full developmental potential— produced a compensatory capacity for growth in their offspring. Typically, evidence of heterosis materialized in the first generation of hybrids, a hypothesis seemingly borne out by fieldwork in Jamaica. Though hybrid vigor was readily apparent in nature—especially in varieties of maize and cotton— observers doubted whether the same held true for humans.[106] Drawing on his studies in Jamaica, Davenport began a correspondence with the efficiency consultant firm of Harrington Emerson in the late 1920s. During an extensive correspondence, Davenport and Emerson mused on the potential benefits of race crossing. Emerson argued, "As with seedlings, extraordinary diversity, some being far above the parent stocks in excellence often occurs." Emerson—a white man—offered a rare endorsement of race mixing: "The line of the Dumas in three generations showed what an infusion of Negro blood could do." Davenport and Emerson agreed that under "proper supervised breeding" in specific environments, hybrids could possibly exhibit "greater adaptive social and industrial powers than their pure-bred ancestors."[107] Employing prewar models of eugenics, Emerson argued that racial traits and tendencies resided in the genes: "The pure Aryan as I know him is essentially coarse minded, his blasphemy is hideous, his obscenity disgusting, his drunkenness brutal. He might be benefited by a blend with a race in which blasphemy is pleasantly familiar, obscenity only piquant, and intoxication delightful."[108] However, this was a minority view among contemporary social scientists and represented a mere, albeit intriguing, detour in Davenport's generally consistent biological worldview.

Davenport's efforts to establish a transnational geography of racial hybridity disabused him of the idea that racial hybrids possessed a kind of "beneficial adaptability." The basal metabolism records—the most accurate index of vital capacity—indicated that "Browns" registered slightly below both blacks and whites. Davenport noted that while mulattos tested between whites and blacks in terms of "physical proportions and mental capacity," an "excessive number seemed not to be able to utilize their native [read: white] endowment."[109] Jamaica's racial demographics—the "undiluted Africanisms of its Negroes" and its small yet "pure European" population—meant that its mixed population comprised a clear mix of "warring identities" entirely incompatible with one another in social and or biological terms. Davenport and Steggerda argued that the insidious effects of race mixing transcended national boundaries—framing their research as incontrovertible evidence for the need to maintain and expand social and legislative prohibitions against race mixing stateside. Both men hoped to prevent the U.S. falling victim to the warring identities of hybridity that had seemingly devitalized and subsequently impoverished many of their southern neighbors.[110]

NRC efforts to construct a body of knowledge on the New Negro type at home and abroad were predicated on the notion that racial homogeneity and bodily integrity were essential to national health—that is, fit peoples informed fit nations. NRC anthropologists retained much of prewar taxonomies in delineating the Negro type but generally eschewed prior hierarchical models of racial stratification. Abram Harris of the NUL noted, "The apologetic school of American race relations considers the social distance between white and black Americans as conforming to a natural order of preordained and inescapable physical, mental, and moral differences." The Negro type of the postwar imagination was defined as much by difference as by degree. Based on wartime evaluations postwar social scientists anticipated that "homogeneity in occupation is directly related to the physical traits of groups."[111] The disproportionate clustering of certain groups in certain occupations was due to the fact that specific peoples were physiologically fitted for specific kinds of work. Therefore the productions of models of racial difference—as opposed to strict hierarchical models of race—were seen as instrumental to the development of the postwar labor economy.

The impetus to mediate the nation's labor economy through racial models of "fit" and "unfit" bodies received legislative sanction in 1924 with the passage of the infamous Immigration Restriction Act and Virginia's Anti-Miscegenation Act, which prohibited inter-racial marriage and provided for the sterilization of the "unfit."[112] Whereas the former was responding to a racial landscape from which the Negro was largely absent, the latter directly addressed postwar fears regarding the emergence of the New Negro. Yet passage of both these acts—though hailed at the time by their supporters as the culmination of a long crusade—actually marked the beginning of the end of eugenically informed social policy: a rearguard action by embattled white nativists. Both statues drew heavily on wartime mental and physical testing in their respective attempts to impose racial quotas on immigration and criminalize interracial sex.[113] Both acts were also predicated on the assumption, cultivated in no small part by the congressional testimonies of eugenicists such as Harry Laughlin, that race and racial difference were real and that only through the cultivation and preservation of the "fittest types" could the nation survive.[114] To counter the "rising tide" of colored labor, the disparate so-called "near white" races of southern, central, and eastern Europe were fused with native-born whites into a common Caucasian identity—making able bodied whiteness a precondition of national health.[115] Melville Herskovits confirmed eugenicists' worst fears, arguing that "the very term 'Negro' is social rather than racial [and, as in the United States,] means 'not all white'"—bedeviling those who wished to reify the corporeal constitution of the color line.[116] Foregrounding working bodies of color as categories of analysis reveals these legislative practices of racial control—not as reactionary aberrations but as constitutive of an American industrial modernity predicated on the discipline and control of "fit" and "unfit" laboring bodies along racial lines.

However, this utopian vision of racial labor control through strict quantification was complicated by NRC anthropologists' experiences in the West Indies, which forced them to rethink the dimensions of the New Negro type. In Jamaica, race mixture seemingly led to "social and occupational stratification based on a degree of white blood which interferes with racial solidarity" across the racial divide.[117] Observers felt that because biological imperatives—the inborn association between like and like—were not driving peoples to self-segregate along racial lines, segregationist legislation was necessary. Anticipating the messy cultural pluralism of the 1930s, Herskovits characterized the faltering hereditarian-engendered biracialism of the postwar era not as white versus black, but as a race to the bottom to determine who was definitively not white and who, therefore, was precluded from the benefits of white privilege.[118]

Beyond the 1924 Immigration Restriction Act and Virginia's Anti-Miscegenation Act, the regulatory initiatives of the Progressive Era, and their draconian wartime manifestations, sapped the public desire for greater government control over social policy. In the spring of 1929, just prior to the publication of *Race Crossing in Jamaica*, the Committee on the American Negro disbanded. In March, Chairman R. J. Terry presented his resignation to Davenport and Boas on the pretext that "someone in the east who is in closer touch with the foundations will assume this function." In reality, the committee had ceased functioning by the fall of 1928. Like its predecessors, the CRC and CSPHM, the Committee on the American Negro failed to secure adequate funding and maintain any kind of institutional stability. The various studies undertaken by its members, the majority of which remained incomplete, were often conducted under the auspices of their respectively better endowed academic or philanthropic foundations.[119] The NRC's drive to create a clearing-house of racial knowledge and develop a racial science—one that was defined by method and oriented to racial labor control—floundered on the shoals of postwar anti-statism. Despite the NRC's efforts to link social biology to national efficiency, and racial form to racial function, there was little aptitude on the part of lawmakers and the public to fund or facilitate the necessary studies.

From the black belt of Dixie to the cane fields of Jamaica to the law courts of Virginia, postwar NRC anthropologists saw mixed race peoples as unsettling avatars of modernity. Previous social scientific assessments of racial labor fitness had been exercises in mutual negation—aimed at determining the work races could not, or would not, do in relation to other races. Social scientists hoped that an evaluation of racially "in between peoples" would allow for a more definitive appraisal of the respective laboring fitness of the New Negro and Caucasian. The logic of Jim Crow led anthropologists to posit the mulatto as a literal mix of black and white types, rather than as a new racial type. Stoddard described "racial mongrels" as those "for whom in every cell of their bodies is the battleground of jarring heredities."[120] In the early 1930s, Otto Klineberg, a graduate of McGill and professor of psychology at Columbia, drew on wartime studies to compare southern and northern "pure and mixed" African Americans "with reference to speed and accuracy of mental and physical performance to determine whether or not there was a racial norm independent of cultural envoir." However, like many NRC studies, Klineberg's faltered from lack of financial resources.[121] For observers across the color line, the "race problem," or more specifically the "Negro problem," was a matter of social biology best mediated through the mulatto, but whose ultimate resolution necessitated the elimination of hybridity in all its forms.

The war and wartime testing reshaped narratives and models of black labor fitness in three significant ways: Wartime migration and military service definitively established the Negro as a factor in industrial civilization. Rapid migration and urbanization were leading to mental and physical changes on African American workers. And national efficiency was linked to racial integrity through transnational discourses of race crossing. Postwar social scientists and their legislative allies constantly despaired over the failure of nature to self correct itself to fit their desired racial ends. For many observers, the very persistence of "unnatural" practices like race crossing was a prime example of the need for scientific intervention in "natural" evolutionary processes. NRC social scientists' persistent and furtive attempts to define and quantify a new Negro type demonstrated that the triumph of whiteness was far from assured by even its most ardent ideological and legislative architects. As early as 1915, DuBois, in "The African Roots of War," observed that one of the unintended consequences of global capitalism was that white-on-white violence on the world stage exposed the dark side of progress and undermined the racial supremacy of progressive industrial nations.[122] The NRC was committed to rebuilding the racial knowledge that had sustained these prewar networks of white supremacy.

The efforts of NRC social scientists to quantify and delineate black labor fitness facilitated the concurrent devaluation of black labor in industry and contributed to blacks' increasing marginalization from the era's labor economy. To paraphrase E. P. Thompson, the New Negro was "present at its own making," a product of processes and relationships—war, migration, and urbanization—which "owed as much to agency as to conditioning."[123] Consequently, new state-sanctioned efforts to link blacks' lack of labor capacity in clear corporeal terms fractured interracial class solidarity. Abram Harris cited the persistence of this "color-caste feeling" for the failure of a logic which teaches that "the ultimate interests of socially disadvantaged whites and blacks are more coincidental than that of white capitalists and white wage earners" to take hold among American workers. This paradox was neatly summarized by the anthropologist Arthur H. Fauset, who claimed, "the New Negro had been in America for a long time, yet everyone had grown so used to seeing Negros that practically no one discovered that differences were taking place under our very eyes."[124] The black philosopher Alain Locke opined, "In the last decade something beyond the watch and guard of statistics has happened in the life of the American Negro and the three norms who have traditionally presided over the 'Negro problem' have a changeling in their laps. The Sociologist, the Philanthropist, the Race Leader are not unaware of the New Negro, but they are at a loss to account for him. He simply cannot be swathed in their formulae."[125] Whereas prewar race theorists had posited culture or race "traits and tendencies" as the basis for biological racial difference, wartime and postwar observers tried to invert this model by substituting biology as an index for racial character or pathology. The futility of this process was on display in postwar studies of race mixing. Wartime testing unwittingly accelerated the decoupling of race from biology—presaging more culturally informed models of racial difference.[126] By the interwar years, war, migration, urbanization, and a perceived rise in race mixing had transformed the Negro into a physically present, yet deeply ambiguous, national figure to social scientists across the color line. Notwithstanding debates regarding the contours and dimensions of the New Negro what was clear was that the Negro—whether in biological or social terms—had become an object of mainstream social scientific inquiry, a problem in perpetual need of a solution.

Epilogue

Invisible Men: The Afterlives of the Negro Problem in American Racial Thought

"The supreme fact of mechanical civilization is that you become a part of it, or get sloughed off (under) . . . A few generations from now, the Negro will still be dark, and a portion of his psychology will spring from this fact, but in all else he will be a conformist to the general outlines of American civilization, or of American chaos."
—Jean Toomer (1923)

"Can a people live and develop for over three hundred years simply by reacting? Are American Negroes simply the creation of white men, or have they at least helped to create themselves out of what they found around them?"
—Ralph Ellison (1964)

Since emancipation the African American experience has been animated by the tension between assimilation and segregation within American society. Though the terminology may have changed—*negro, colored, black, African American*—the question remains: Can African Americans be effectively reconciled to the social, economic, political, and cultural imperatives of American capitalism and democracy? As a racial minority in a historically majority white—or perhaps more accurately "non-black"—nation, the role of blacks in the republican body politic has been consistently conceived and articulated in problematic terms. This conception of blackness as "problem," as inherently inimical or at best incidental to the imperatives of modern American capitalism has been mediated through the corporeal metaphor of the black working type for much of the twentieth century. Indeed, the production of racial difference has been central to the development and maintenance of modern capitalist labor economies in the United States and throughout the globe.[1] Robert Miles argues that racism functions as part of the capitalist labor process not merely as an ideology, but as a means to shape the way work is organized and exploited in historically contingent circumstances. And as Paul Gilroy reminds us, "there is no racism in general"—only an ideological imperative born of specific time, place, and power structure(s).[2] The working black body of the progressive imagination is thus revealed as both a function and an agent of blacks' transition into turn-of-the-century industrial modernity.

Embedded in the sediment of Progressive Era thought, the policies and practices used to define a laboring Negro type provide a fossil record of how capitalism, war, and the modern nation state worked to produce racial and labor hierarchies at the high tide of American industrial modernity. World War I mobilized African Americans for the work of war and organized social scientists to develop new methods of measuring racial labor fitness to potentially transform rural southern black migrants into modern worker-soldiers under the auspices of a nascent military industrial complex. The imperatives of the wartime state intensified the proliferation of various regulatory institutions such as the draft and vocational rehabilitation for the surveillance and discipline of the black working body. Positing the working black body as a site of inquiry, discipline, and knowledge production, fields such as sociology, anthropometry, vocational rehabilitation, and anthropology failed to create new and lasting racial labor taxonomies of racial labor fitness yet affirmed race as an organizing principle of American labor economy. Wartime testing merely exacerbated and refined the "liberal Enlightenment commitment to categorizing, classifying, and controlling." Rather than representing a breakdown of Western civilization, the war laid bare its governing racial epistemologies in distinctly corporal terms.[3]

Regardless of whether the "Negro or race problem" was constituted in biological or culturist terms, the *need* for a metaphorical Negro type has remained remarkably persistent from the Gilded Age imagery of Ben Bailey, to the Negro of World War I testing, the pathological Negro of the Civil Rights era, to the criminalized black bodies of today. The author Richard Wright described the Negro as "America's metaphor"—one that is intimately related to the legacy of slavery, but ultimately without fixed cultural or biological attributes. For Wright, "the word Negro in America means something not racial or biological, but something purely social, something made in the United States."[4] In his landmark study of American race relations, *An American Dilemma*, the Swedish economist Gunnar Myrdal remarked that "the Negro has to be defined according to social usage, and his African ancestry and physical characteristics are fixed to his person much more ineffaceably then the yellow star is fixed to the Jew."[5] Herein lay the conundrum of the Negro as a decidedly embodied presence yet ever shifting social construct—a "changing same" of racial identity.[6] Indeed the persistence of race and racial difference as metaphor can be attributed to its infinitely malleable nature.[7] For Sundiata Keita Cha-Jua, "racial capitalism has been undying and constantly changing, shedding its old skin as Douglass said it would and reappearing in ever newer forms: slavery, sharecropping, proletarianization, and the labor marginalization" of contemporary postindustrial globalization.[8]

Racial hierarchies have historically been constitutive in the development of the American capitalist state and will invariably mutate to accommodate shifting racial demographics—America is changing from a traditional white-black binary to a brown or beige nation, and the so-called racial minorities will soon constitute the majority population. Yet despite these demographic shifts, the immutable blackness of African Americans as metaphor and objects of social scientific inquiry remains.[9] Any and all postmodern theories of race as metaphor or social construct must be incorporated into the structural framework of American capitalism. As Theodore Koditschek notes, "whether or not we want to call our epoch 'postmodern,' it is still very much capitalist."[10] Notwithstanding pretentions to a postmodern or post-racial present, race still functions as a concrete means to define and create social inequality within these systems. Despite the growth of a robust black middle class, greater representation of blacks in corporate and political America, and the election of the first African American president Barack Obama in 2008, black and white Americans continue to be separated by vastly unequal access to jobs, education, capital, home ownership, and health care. Perhaps the most significant evidence of this divide can be found in the criminal justice system. Though blacks account for only approximately 13% of the national population, they comprise almost 60% of the nation's prison's population and are three times more likely to be imprisoned than Latinos and seven times more likely than whites. Shockingly, there are more African Americans incarcerated, on probation, or on parole than were enslaved in the United States in 1850.[11] Despite the growth of a substantial black middle class, millions of blacks still reside in the nation's inner cities and are effectively cut off from the mainstream labor economy—instead, they are left toiling in menial service industry jobs or as agents in the violent drug trade. Neoliberal economic policies, which only serve to further increase the mobility of capital, have led to rampant deindustrialization—reducing many American inner cities to veritable wastelands populated by poor, elderly, disabled peoples of color, totally cut off from mainstream labor economies. For many observers across the political spectrum, the very presence of black people and black bodies within the body politic ensures the persistence of racism given their understanding of blacks' seemingly inherently problematic nature.[12]

Despite differences in method and historical context, the consistent framing of blacks and black bodies as mere objects of inquiry has made black humanity disposable in the calculus of the industrial past and the postindustrial present. A little over a century ago, Kelly Miller enjoined his fellow African Americans not to fall victim to prevailing "expert" opinion regarding the Negro's inherent inferiority in impressively prescient terms: "He [the Negro] does not labor under a destiny of death from which there is no escape. It is a *condition* and not a *theory* that confronts him."[13] Some two generations later, the author Ralph Ellison argued for a black humanity that was more than the sum of its allegedly pathological parts. Perpetual attempts to "solve the Negro" buried living, breathing people under an avalanche of theories, methodologies, and numbers—rendering them little more than objects of pity or derision.[14] Yet Ellison's appeal for a full and insistent black humanity that would make visible a heretofore largely invisible people in the eyes of most Americans still resonates in our ostensibly post-ideological, post-racial, neoliberal age. Recent appeals to social justice and racial equality, such as the Black Lives Matter movement, reveal with depressing regularity the need to defend the autonomy of black lives, bodies, and souls to wider American society.[15] Yet even a cursory glance of American history reveals the ways in which the devaluation, subjugation, and destruction of black bodies—from the auction block to the prison cell—has informed the development of American political and labor economies.

Reimagining race and racial difference as *needs* born of specific historical contexts rather than *facts* divests them of their seemingly natural, inevitable, and ahistorical character. Throughout history these imperatives have been predicated on the simple axiom that black bodies, black health, and black lives do not, and in fact cannot, matter, for the many headed hydra that is American capitalism to function. Given that race and racial difference have primarily been normalized through the body, it is through a corporeal lens that these constructions must be deconstructed and challenged. Positing working black bodies as objects of inquiry, social constructions, and above all economic necessities born of historical contingencies facilitates a critique of the allegedly "color blind" logic of market capitalism. Historicizing the laboring black body as an imperative of national and global labor economies reveals race and racial inequality to be constitutive to the development and rationalization of past, present and future cultures of American capitalisms that ensnare us all.[16]

NOTES

INTRODUCTION

1 *Philadelphia Record*, Morning Edition, March 23, 1885. On black boxing in late nineteenth-century Philadelphia, see Roger Lane, *The Roots of Violence in Black Philadelphia, 1860–1900* (Cambridge, MA: Harvard University Press, 1986), 118–119.

2 *Philadelphia Record*, Morning Edition, March 23, 1885; *Philadelphia Record*, Morning Edition, April 14, 1885.

3 The color line in boxing at this time was informal and inconsistent in its application. Theresa Runstedtler notes that "profit motives and consumer demand foiled attempts to maintain a rigid color line." In later years, modern technologies such as radio and film would prevent its potential architects from restricting the wider public's access to the triumphs of black boxers, most notable being those of Jack Johnson, the world heavyweight champion from 1908 to 1915 and the scourge of global white supremacy. Runstedtler, *Jack Johnson, Rebel Sojourner*, 18.

4 Though August 1883 signaled the formalization of Muybridge's Motion Studies at Penn, he had spent much of the previous winter and spring in Philadelphia giving a series of lectures on Animal Motion at the Pennsylvania Academy of the Fine Arts, the Franklin Institute, and the Academy of Music, sites that evinced Muybridge's artistic understandings of his work while also bringing him to the attention of a diverse and influential group of faculty and administrators at Penn. Gordon, "Prestige, Professionalism, and the Paradox," 86–87.

5 Eadweard Muybridge, "Muybridge at Penn," "Animal Locomotion," Prospectus and Catalog of Plates (1887) UPT 50 M993, Box 62, Folder 2, Eadweard Muybridge Papers, University Archives, University of Pennsylvania; Brown, "Racialising the Virile Body," 627–628.

6 Muybridge, "Muybridge at Penn," Draft of "*Animal Locomotion: An Electro-Photographic Investigation of Consecutive Phases of Animal Movements*" (1887) UPT 50 M993, Box 62, Folder 2, Muybridge Papers, University Archives, University of Pennsylvania.

7 Brown, "Racialising the Virile Body," 637.

8 Ibid., 637–638.

9 On the grid as a means of racial knowledge production, Elspeth Brown notes "it is as if the non-white 'other' cannot be understood, scientifically, without the anthropometric grid, a technology for mapping racial difference" in stark physiological terms. Brown, "Racialising the Virile Body," 637–638.

10 Wayne, *Imagining Black America*, 3–5.
11 During the 1890s, lynching claimed some 139 lives each year, 75% of which were black. Litwack, *Trouble in Mind*, 436–437. For blacks'—specifically black women's—appropriation of racial violence into narratives of corporeal resistance, see Williams, *They Left Great Marks on Me*.
12 Although the racial logic of the day, specifically the one-drop rule, classified Bailey as a black man, continual reference to his "mulatto" identity complicates his status as a representative and definable black body and speaks to the era's often-elastic definition of blackness—an elasticity that Muybridge's framing of Bailey against the grid would help to undermine and that would presage the hardening of racial boundaries in the scientific and popular imaginary in turn-of-the-century America. Brown, "Racialising the Virile Body," 637–638.
13 On the socio-cultural construction of neurasthenia in Progressive Era America and its racial and gender dynamics, see Bederman, *Manliness and Civilization*.
14 On the civilized and savage dichotomy that informed assessments of turn-of-the-century white and black laboring bodies, see Bender, *American Abyss*, 88–90. For insight on how emergent forms of professional expertise coalesced in the modern university and informed public debates regarding the definition and limits of propriety (specifically nudity) in the public sphere, see Gordon, "Prestige, Professionalism and the Paradox," 81–82. On "primeval" blackness as atavism, see Seitler, *Atavistic Tendencies*, 63–64.
15 In Italy, degeneration was embodied in the malformed body of the criminal; in France, the mentally ill were seen as the greatest threat to national health; meanwhile, in England, it was the specter of the working poor that constituted a cancer upon the national body while in America it was the Negro who was seen as the embodiment of degeneration. Nye, "Degeneration, Neurasthenia."
16 Much of the theoretical and conceptual basis of this work—positing the working body as a site and agent of social and knowledge production—draws on Anson Rabinbach's work on the *human motor*, a metaphor of work and energy that provided nineteenth-century European thinkers with a new scientific, cultural framework for making sense of society. For Rabinbach, the metaphor of the human motor "translated revolutionary scientific discoveries about physical nature into a new vision of social modernity." Building on this formulation I contend that through the metaphor of the Negro working type, progressives sought to maintain blacks' traditional social marginalization among increased occupational and interregional mobility, to naturalize or embody racial hierarchies—from the factory floor to the trenches of France. Rabinbach, *The Human Motor*.
17 DuBois, quoted in Lewis, *W. E. B. DuBois*, 171. For DuBois and many of his peers across the social and natural sciences in thrall to Hegelian notions of cultural determinism, culture preceded biology in the process of racial development, as outlined in DuBois's "Conservation of Races" (1897).
18 Bender, *American Abyss*, 21–22; Banta, *Taylored Lives*, 4.
19 Painter, *The History of White People*, 1–3.

20 Thompson, *Witness Against the Beast*, xii, xix; Williams, *Marxism and Literature*, 55–71.

21 Stepan, *The Idea of Race in Science*, xvi.

22 Stanley, *Bondage to Contract*, 95–97.

23 Critics such as E. P. Thompson challenged this determinist model of working class development by positing an active process of class making that owed as much to the dynamic responses of workers as it did to the impact of industrialists and machines noting "the working class did not rise like the sun at the appointed time. It was present at its own making." Thompson, *The Making of the English Working Class*, 9–14. Wheen, *Marx's 'Das Kapital'*; Braverman, *Labor and Monopoly Capital*; Montgomery, *Workers' Control in America*.

24 The seminal works on black proletarianization remain Trotter, *Black Milwaukee*, esp. 276–277; Trotter, *The Great Migration in Historical Perspective*.

25 Examples of migration narratives include Gottlieb, *Making Their Own Way*; Grossman, *Land of Hope*; Phillips, *Alabama North*; Wilkerson, *The Warmth of Other Suns*. For examples of the focus on black agency engendered in migration and war, see Kelley, *Race Rebels*.

26 On the progressive imperatives of the New South, see Blackmon, *Slavery by Another Name*; Cell, *The Highest Stage of White Supremacy*; Liechtenstein, *Twice the Work of Free Labor*; and Ayers, *The Promise of the New South*. For a post-Progressive Era perspective on the persistence of these models of racial labor control, see Cobb, *The Most Southern Place on Earth*.

27 Ray S. Baker, "A Statesmen of the Negro Problem," *World's Work* (Summer 1916). On black soldiers as avatars of modernity, see Williams, *Torchbearers of Democracy*.

28 Analysis of black working class agency must be tempered by an acknowledgement of the overwhelming power of the state, especially in wartime, in sanctioning categories of race and racial labor fitness. Historians of race and labor need to heed the admonishment of scholars to "bring the state back in" to better determine how racial labor hierarchies are embodied on both the shop floor and the battlefield. Evans, et al., *Bringing the State Back In*.

29 On racial corporality as social knowledge conduits, see Schilling, *The Body and Social Theory*, 16–17; Wailoo, *Dying in the City of the Blues*. On past and present attempts to mediate racial progress through corporally-infused narratives of health, see Downs, *Sick from Freedom*.

30 Boris and Baron, "The Body as a Useful Category," 23–26, note that "bodies are both constituted by and constitutive of the workplace they inhabit and the racialized and gendered class relations which work both expresses and creates." Moreover, "racial bodies are cultural productions, constituted through an interpretive process which often masks the social struggles that went into their making."

31 I build on Catherine Kudlick's call for disability as a category of analysis when delineating how social difference has been embodied throughout history. Kudlick's argument for "why we need another 'other'" is especially useful for

charting the way in which race and labor hierarchies were understood in visceral, aesthetic and everyday terms. Kudlick, "Disability History." Also see Longmore and Umansky, *The New Disability History*. For insights on the relationship between citizenship as a function and agent of conscription, see Christopher Capozzola, *Uncle Sam Wants You: World War I and the Making of the Modern American Citizen* (New York: Oxford University Press, 2010).

32 Foucault, *History of Sexuality*, 139–144.

33 On blacks' uses of lynching and racial violence as corporeal repositories of resistance, see Williams, *They Left Great Marks on Me*.

34 On quantification—along corporeal lines or otherwise—as both constitutive and an imperative of the modern nation state, see Scott, *Gender and the Politics of History*, 113–138. Bailey's last recorded bout on March 6, 1890 was a loss by knockout to a Mike Boden in Philadelphia. Over the span of a five-year career from 1885 to 1890, Bailey carved out a decidedly unspectacular record of four wins (two by knockout), five losses (two by knockout), and two draws. "Ben Bailey," BoxRec.com, http://boxrec.com/list_bouts.php?human_ id=565556&cat=boxer. Accessed March 20, 2015.

35 Brantlinger, *Dark Vanishings*, 2–3.

36 Frederick L. Hoffman, "The Practical Use of Vital Statistics," *Boston Medical and Surgical Journal* 143, no. 26 (December 1900), 653; Porter, *Trust in Numbers*, 11, 45–53; Porter, "Statistical Utopianism," 210–227.

37 Louis Menard, *The Metaphysical Club: A Story of Ideas in America* (New York: Farrar, Straus and Giroux, 2002), xxi, defines the Progressive Era as one in which "ideas mattered" while remaining committed to the notion that ideas should never ossify into ideologies. Lorenzo Fioramonti's reading of Max Weber on bureaucracy is especially instructive here. Fioramonti, *How Numbers Rule the World*, 20, argues that Weber sees the essence of bureaucracy as the power of technology, which leads to the marginalization of all irrational and emotional elements associated with political factors, that is, human factors (i.e. race) that escape the precision of calculation.

38 Degler, *In Search of Human Nature*, viii–ix, 75–78; Bender, *American Abyss*, 2–10.

39 Walter Lippman, "Negro Migration," *New Republic*, July 1, 1916.

40 Biddle, "Military History, Democracy," 1143–1145.

41 Banta, *Taylored Lives*, 4–6. Debates persist over whether the various social, economic and cultural attempts to negotiate what James Livingstone describes as a shift from proprietary to corporate capitalism evinced a retrenchment of the corporatist status quo or produced new ruling structures and values. Livingstone, *Pragmatism and the Political Economy*, xxiii–xxv.

42 Rabinbach, *The Human Motor*, 4–8.

43 For analysis of transnational networks of Progressive reform, see Rodgers, *Atlantic Crossings*, 279.

44 Gains, *Uplifting the Race*. DuBois later characterized his unsuccessful pursuit of a military commission as an attempt "to increase the race's knowledge capital at a time of great crisis." Lewis, *W. E. B. DuBois*, 552–560.

45 On the links between military and labor history, see Way, "Class and the Common Soldier," 455–481; Way, "Rebellion of the Regulars," 761–792; Stanfield, "The Negro Problem," 188; Wilson, *The Segregated Scholars*, 5.

46 Findings collected in Davenport and Love, *Physical Examination of the First Million Draft Recruits*; *Defects Found in Drafted Men*; and *Army Anthropology*.

47 Holmes to Hrdlicka, Report of the Committee on Anthropology, *Proceedings of the National Academy of Sciences* (1918). W. E. B. DuBois was the foremost black practitioner of anthropometry or biometric science. For insight into how DuBois engaged with this scholarship, used it to bind his scientific and literary work, and reconfigured it as a means of uplift, see Farland, "W. E. B. DuBois, Anthropometric Science," 1017–1044.

48 On race and racial division—specifically blackness—as a *function* of Western/American modernity, see Cornel West, "A Geneology of Modern Racism," in *Prophesy Deliverance: An Afro-American Revolutionary Christianity* (New York: Westminster John Knox Press, 1982), 154, 162. On race and labor management as constitutive to the development of modern American labor economies, see Roediger and Esch, *The Production of Difference*.

49 Gerber, *Disabled Veterans in History*, 4; Gelber, "A 'Hard-Boiled Order,'" 161–180. For insights into rehabilitation as a mechanism of imperial racial control, see Elizabeth West, "Divine Fragments: Even India has Begun to Salvage its Man Power," *Carry On* 1, no. 3 (1918), 22–25; and Frader, "From Muscles to Nerves," 123–147.

50 *Literary Digest*, June 14, 1919, 23. Nonetheless, just as mainstream social sciences began to abandon hereditarianism, leading African American nationalists such as Marcus Garvey began to embrace race purity as a source of racial uplift. See Bay, *The White Image in the Black Mind*, 215–217.

51 "The Determination of Racial Relationships by Means of Blood," *Journal of the American Medical Association* 73 (December 27, 1919), 1941–1942; Lippman, "Negro Migration," *New Republic*, July 1, 1916; Adas, *Dominance by Design*, 281.

52 Visweswaran, "Race and the Culture of Anthropology," 70, argues that the "attempt to expunge race from social science by assigning it to biology as Boas and his students did, helped to legitimate the scientific study of race, thereby fuelling the machine of scientific racism."

53 Madison Grant, *The Passing of the Great Race* (New York: Scribner, 1916); Stoddard, *The Rising Tide of Color*.

54 DuBois, *Darkwater*, 32.

55 The pushback against eugenic thinking has often been characterized as coinciding with the rise of New Deal pluralism, abetted by new models of cultural relativism developed by the anthropologist Franz Boas and his students. For many scholars, eugenic models of race and racial difference were ultimately discredited with the rise of Nazism and the horrors of Auschwitz. Daniel Kelves notes that while the "barbarousness of Nazi policies eventually provoked a powerful anti-eugenic reaction . . . this reaction obscured a deeper historical reality: many thoughtful members of the American public had already recognized that a great deal was

wrong with mainline eugenics" and the general urge to mediate social policy in strict biological terms. Baker, *From Savage to Negro*; Pickens, *Eugenics and the Progressives*; Kelves, *In the Name of Eugenics*, 118–119; Bender, *American Abyss*, 243–246.

56 DuBois, *The Souls of Black Folk*, 213–215.

57 Painter, *The History of White People*, 1; Banta, *Taylored Lives*, 29; Zimmerman, *Alabama in Africa*, 40.

58 For additional examples of works on the "Negro problem," see Jabez L. Curry, "The Negro Question," *Popular Science Monthly* 55 (1899); and LeConte, "The Race Problem in the South." Theories of degeneration also infused a transnational reform discourse. See Nye, "Degeneration, Neurasthenia," 51–69; Nye, *Crime, Madness and Politics*; and Pick, *Faces of Degeneration*. Lears, *Rebirth of a Nation*, outlines the counter discourse of regeneration which animated Progressive Era thought.

59 Downs, *Sick from Freedom*, 168–169.

60 On efforts to link work and empire to the development of American modernity, see Bender and Lipman, *Making the Empire Work*.

61 Painter, *The History of White People*, x; James Bryce, "Thoughts on the Negro Problem," *North American Review* 153 (December 1891), 659–660.

62 See Stanley, *Bondage to Contract*, 95–97. Baldwin, *Chicago's New Negroes*, 247, notes that during this period, the labor management practices of "racial reasoning and regulation shifted from the religious tales of Ham to the bio-cultural social sciences."

63 Charles Rosenberg, "Framing Disease," in Rosenberg and Golden, *Framing Disease: Studies in Cultural History*, xiii–xxvi.

CHAPTER 1. MORTALITY AS THE LIFE STORY OF A PEOPLE

1 Frederick L. Hoffman, "Memoir" (unpublished) Frederick L. Hoffman Papers, Box 9, Butler Library Rare Book and Manuscripts, Columbia University, 72–73; Ella Hoffman, "Biography of Frederick L. Hoffman" (unpublished) Vol. 4, Hoffman Papers, Box 31, Butler Library Rare Book and Manuscripts, Columbia University, 117.

2 Ibid., 117–118.

3 Seitler, *Atavistic Tendencies*, 63–66.

4 Hampton Institute had been directly modeled on Hawaii's Hilo Boarding School, founded a generation prior by Samuel's father, Richard Armstrong to bring "thrift and industry" to the islands natives. "The Negro and the Polynesian have many striking similarities," observed the younger Armstrong. "Of both it is true that not mere ignorance, but deficiency of character is the chief difficulty, and that to build up character is the true objective point of education." Gary Okihiro, *Island World: A History of Hawai'i and the United States* (Berkeley: University of California Press, 2008), 114–115. For more on the Hampton, Hawaiian connections, see Engs, *Educating the Disfranchised and Disinherited*.

5 Okihiro, *Island World*, 114–116. Alexander Saxton argues that the sentimental domestication of racial "others"—the noble Indian, for example—into popular culture became possible only through the discourse of degeneracy and extinction because it alleviated the racial anxiety of white Americans. Saxton, *The Rise and Fall of the White Republic*, 342–345. See also Deloria, *Playing Indian*.

6 Hoffman, *Race Traits*; Hoffman, "Vital Statistics of the Negro."

7 Glenn, "Postmodernism," 6, 131–143.

8 For analysis of the transnational models of racial labor knowledge and black vocational uplift, see Zimmerman, *Alabama in Africa*, 40.

9 Degler, *In Search of Human Nature*, viii–ix, 75–78; Bender, *American Abyss*, 2–10; Banta, *Taylored Lives*, 4.

10 Bender, *American Abyss*, 18–19; Hoffman, "The Practical Use of Vital Statistics," 653.

11 Livingstone, *Pragmatism and the Political Economy*, xxiii–xxv.

12 Porter, "Statistical Utopianism," 210–227; "Life Insurance, Medical Testing and the Management of Mortality," in Lorraine Daston, ed., *Biographies of Scientific Objects* (Chicago: University of Chicago Press, 2000), 226.

13 Scott, *Gender and the Politics of History*, 113–138.

14 Kelves, *In the Name of Eugenics*, ix, 13–14. See also Hacking, *The Social Construction of What?*.

15 Fioramonti, *How Numbers Rule the World*, 15.

16 On antebellum black bodies as both embodied and imagined capital, see Johnson, *Soul by Soul*.

17 Painter, *The History of White People*, 245.

18 Because blacks faced discrimination in housing markets and had little capital for down payments, a relatively high proportion of their wealth was held in personal property—such as clothes and furniture—rather than real estate, which make it somewhat difficult to gauge black wealth at the turn of the century. Moreover, the situation differed for urban and rural blacks: in Atlanta blacks possessed only $37,000 worth of real estate in 1869 an amount which had grown to $855,561 by 1900. In contrast, in 1900 black farm ownership was only 8% in the black belt in compared with 54% of whites. Ayers, *The Promise of the New South*, 70, 429, 449–450.

19 Brantlinger, *Dark Vanishings*, 2–3.

20 Rodgers, *Atlantic Crossings*, 209–217.

21 DuBois, *The Philadelphia Negro*, 225; Weare, *Black Business in the New South*, 7.

22 Zelizer, *Morals and Markets*, xii. From its inception, life insurance was opposed by various religious and secular thinkers who objected to the commodification of sacred and personal entities and experiences like bodies, life and, death. In *The Economic and Philosophic Manuscripts*, the young Karl Marx deplored the devaluing of human life beyond the all-encompassing cash nexus. Marx cited labor, prostitution, and slavery as prime examples of the degrading, alienating process of capitalist commodification. Marx, "The Economic and Philosophic Manuscripts," 131–146.

23 Lears, *Rebirth of a Nation*, 93.

24 Wolff, "The Myth of the Actuary," 84–91.

25 Ibid.

26 Hoffman, *History of Prudential*, 16, 153, 185.

27 Wolff, "The Myth of the Actuary," 88; Hoffman, *History of Prudential*, 185.

28 Francis Sypher, "The Rediscovered Prophet: Frederick L. Hoffman (1865–1946)," Cosmos Club (2000), http://cosmos-club.org/web/journals/2000/sypher.html.

29 Ibid.

30 N. S. Shaler, "The Negro Problem," *Atlantic Monthly*, November 1884, 703; Brantlinger, *Dark Vanishings*, 2–3.

31 Beard, quoted in Thomas Bender, *A Nation among Nations: America's Place in World History* (New York: Hill and Wang, 2006), 247.

32 Bender, *American Abyss*, 119–121.

33 For analysis of minstrelsy as a key cultural discourse of nineteenth-century white supremacy, see Lott, *Love and Theft*.

34 Warren, "Northern Chills, Southern Fevers," 7.

35 Eugene R. Corson, "The Vital Equation of the Colored Race and Its Future in the United States," in *Wilder Quarter-Century Book* (Ithaca, 1893), 123.

36 Bryce, "Thoughts on the Negro Problem," 659–660.

37 Hoffman, *Race Traits*, 6, 148, 312. Previous to 1860, Chinese and Indians were counted as colored; for 1860 and 1890 they were excluded altogether.

38 Ibid., 142.

39 Ibid., 6.

40 Zelizer, *Morals and Markets*, xii.

41 Ibid., 113, 48.

42 Hoffman, *Race Traits*, 28.

43 Of the 431 black babies born in Atlanta for the year 1895, a staggering 194 (or 45%) died before their first birthday; however as deaths were recorded more often then births the real infant mortality rate was likely somewhat less. Litwack, *Trouble in Mind*.

44 Hoffman, *Race Traits*, 64–68, 33–37.

45   Frederickson, *The Black Image in the White Mind*, 238.

46 Litwack, *Trouble in Mind*, 483; Hoffman, *Race Traits*, 17–18. Also see Brian Kelly, "Industrial Sentinels Confront the 'Rabid Faction': Black Elites, Black Workers, and the Labor Question in the Jim Crow South," in Arnesen, *The Black Worker*.

47 Galishoff, "Germs Know No Color Line"; Hoffman, "Vital Statistics of the Negro," 534.

48 Muhammad, *The Condemnation of Blackness*, 5.

49 Hoffman, *Race Traits*, 226, 263–265.

50 Ibid., 263–265.

51 Pick, *Faces of Degeneration*, 109–138; Horn, *The Criminal Body*.

52 Hoffman, *Race Traits*, 218–220.

53 Litwack, *Trouble in Mind*, 436–437; Hoffman, *Race Traits*, 218–225.

54 Litwack, *Trouble in Mind*, 284.
55 Ibid., 92–95.
56 Ibid., 96.
57 Ayers, *The Promise of the New South*, 158. The analogy of blacks as animals was pervasive in nineteenth-century American culture. With the coming of Darwinian theory, ethnologists continued to stress the animal nature of blacks by positing them as the "missing link" between apes and white men. See Haller, *Outcasts from Evolution*; Humphreys, *Intensely Human*.
58 Bay, *The White Image in the Black Mind*, 190–191.
59 Hoffman, *Race Traits*, 95–96.
60 Quoted in Bay, *The White Image in the Black Mind*, 190–191.
61 Hoffman, *Race Traits*, 185–187, 180–182.
62 Ibid., 156.
63 Carter, *The United States of the United Races*, 77–106.
64 LeConte, quoted in Frederickson, *The Black Image in the White Mind*, 246–247.
65 These findings were complicated by the fact that mulattos were not listed in either the 1880 or 1900 Census. Williamson, *New People*, 112.
66 Hoffman, *Race Traits*, 186.
67 *Spectator*, June 1897.
68 Hoffman, *Race Traits*, 140.
69 Smithers, *Science, Sexuality, and Race*, 158–161. Also see Mitchell, *Righteous Propagation*. On the mulatto as a barometer of the racial boundaries of the republican body politic, see Carter, *The United States of the United Races*.
70 Hoffman, *Race Traits*, 140.
71 Rabinbach, *The Human Motor*, 23–25.
72 Ibid.
73 Braun, "Spirometry, Measurement, and Race," 136–137.
74 Gould, *Investigations in the Military*.
75 Hoffman, *Race Traits*, 310–312.
76 Ibid., 183–185.
77 Ibid.
78 Ott, *Fevered Lives*, 12.
79 Ibid., 84.
80 DuBois, *The Philadelphia Negro*, 160.
81 For more on how tuberculosis (TB) contributed to a fracturing of working class whiteness along racial and gendered lines for Jewish immigrants, see Bender, *Sweated Work, Weak Bodies*.
82 Schweik, *The Ugly Laws*, 193.
83 Franklin, *From Slavery to Freedom*, 400–401.
84 Hoffman, *Race Traits*, 70–77.
85 Ibid., 70.
86 Ibid., 160.
87 Bender, *American Abyss*, 43–46.

88 See Anderson, *Colonial Pathologies*; McCallum, *Leonard Wood*.

89 Guterl and Skwiot, "Atlantic and Pacific Crossings," 46; Anderson, *Colonial Pathologies*.

90 For more on quantification—racial and otherwise—as an imperative of empire, see Ian Hacking, "Why Race Still Matters," *Daedalus* 135 (Fall 2006), 113–115.

91 Banta, *Taylored Lives*, 27.

92 M. Dawson, "Review of *Race Traits and Tendencies of the American Negro*," *Publications of the American Statistical Association* 5, no. 35 (September–December 1896).

93 Quoted in Hoffman, "Biography of Frederick L. Hoffman" (unpublished) Vol. 4, Hoffman Papers, Box 31, Butler Library Rare Book and Manuscripts, Columbia University, 120.

94 F. Lamson Scribner, "Review of *Race Traits and Tendencies of the American Negro*," *Science* 5, no. 106 (January 8, 1897), 62–69.

95 W. B. Smith, *The Color Line: A Brief in Behalf of the Unborn* (New York, 1905), 186–187, 190–191.

96 *Dial Magazine*, January 1897, 17.

97 Rogers, *Atlantic Crossings*, 125.

98 DuBois, *Annals of the American Academy of Political and Social Science* (January 1897).

99 On DuBois's "Germanophilia," see Lewis, *W. E. B. DuBois*, 127–130.

100 Reed, *W. E. B. DuBois and American Political Thought*, 32–35; Katz and Sugrue, *W. E. B. DuBois, Race, and the City*.

101 In 1810, there were 1,377,808 blacks in the U.S., a number that had swelled to approximately 7,470,040 in 1890. Kelly Miller, "Review of *Race Traits*," The American Negro Academy, Occasional Papers, no. 1 (1897).

102 Ibid.

103 Ibid.

104 Logan, *The Betrayal of the Negro*, 324.

105 Harry Pace, "The Attitudes of Life Insurance Companies Towards Negroes," *Southern Workman* 57, no. 5 (January 1928).

106 DuBois, quoted in Ayers, *The Promise of the New South*, 430.

107 Ibid.

108 Gaines, *Uplifting the Race*, 14.

109 Frederick Hoffman, "Race Pathology," Hoffman Papers, Box 11, Folder 12, Butler Library Rare Book and Manuscripts, Columbia University.

110 Weare, *Black Business in the New South*, 280–281.

111 Charles Carroll, *The Negro a Beast: or, in the Image of God?* (St. Louis, MO: American Book and Bible House, 1900); W. P. Calhoun, *The Caucasian and the Negro in the United States* (Arno Press, 1902); Smith, *The Color Line*; Robert Shufeldt, *The Negro: A Menace to American Civilization* (Boston: Gorham Press, 1907).

112 Eggleston, *The Ultimate Solution*.

113 Ibid., 124.
114 Barringer, *The American Negro*.
115 Dorr, *Segregation's Science*, 22–24.
116 Hoffman, *Race Traits*, 312–313. See Marion Dawson, "The South and the Negro," *North American Review* 172 (February 1901).
117 Guterl, *The Color of Race*, 117.
118 Hoffman to Dryden, November 2, 1914, Hoffman Papers, Box 13, Folder 11, Butler Library Rare Book and Manuscripts, Columbia University.
119 *Honolulu Commercial Advertiser*, April 1901.
120 Frederick Hoffman, "Travels in Hawaii," Hoffman Papers, Box 13, Butler Library Rare Book and Manuscripts, Columbia University.
121 Ibid.
122 Hoffman, "Race Pathology."
123 Frederick Hoffman, *The Sanitary Progress and Vital Statistics of Hawaii* (Newark: Prudential Life Insurance Company, 1916), 9.
124 Ibid., 9–11.
125 See Gary Okihiro, *Pineapple Culture: A History of the Tropical and Temperate Zones* (Berkeley: University of California Press, 2009).
126 Frederick Hoffman, "Races of Mankind," Hoffman Papers, Box 14, Folder 3, Butler Library Rare Book and Manuscripts, Columbia University.
127 Ibid.
128 Stepan, *The Idea of Race in Science*.
129 Schweik, *The Ugly Laws*, 193.
130 Ibid., 193–194.
131 Booker T. Washington, quoted in Bruinius, *Better for All the World*.
132 Porter, *Trust in Numbers*, 124–126.
133 Wolff, "The Myth of the Actuary," 121. For an examination of the racial epistemologies of statistics, see Tukufu Zuberi, "Deracailyzing Social Statistics: Problems in the Quantification of Race," *Annals* (March 2000), 173.
134 Wolff, "The Myth of the Actuary," 121–122.

CHAPTER 2. THE NEGRO IS PLASTIC

1 *Survey* (July 14, 1917), 331, 333.
2 James Winston, *Holding Aloft the Banner of Ethiopia: Caribbean Radicalism in Early-Twentieth Century America* (New York: Verso, 1999), 96.
3 Wayne, *Imagining Black America*, 44–47.
4 Memo to Post, August 1, 1917, Records of the War Production Board, Record Group (RG) 179, National Archives Building, College Park, MD (NAB II).
5 Ibid.
6 For an analysis of how workers exercised white privilege within the nexus of work and culture, see Lott, *Love and Theft*; Saxton, *The Rise and Fall of the White Republic*; Ignatiev, *How the Irish Became White*; Allen, *The Invention of the White Race*.

7 Stanfield, "The Negro Problem," 188.

8 Despite being the first federal agency since Reconstruction exclusively devoted to the black worker, the DNE is largely absent from African American and Progressive historiography. Until quite recently, much of this ambivalence was due to historians' tendency to interpret the modern African American experience through the narrow lens of Washingtonian accommodation and DuBoisian radicalism. For the former, Haynes's moderate racial liberalism was a corrective to the perceived excesses of the black power movement of the late sixties and early seventies. For scholars committed to detailing the agency and forms of resistance of the black working class, Haynes's brand of bourgeois respectability was seen as irrelevant, or even detrimental, to the cause of black equality. However, recent works by historians Toure Reed and Francille Wilson have helped to re-conceptualize Haynes's commitment to developing black labor expertise and his efforts to instill in migrants an industrial consciousness as complex and nuanced attempts to assert black labor fitness at a high tide of eugenics and Jim Crow. Reed, *Not Alms but Opportunity*; Wilson, *The Segregated Scholars*, 171.

9 Francis Walker, quoted in Painter, *The History of White People*, 212.

10 McKee, *Sociology and the Race Problem*, 29.

11 Baker, *Anthropology and the Racial Politics*, 5.

12 Edward A. Ross, "The Causes of Race Superiority," *Annals of the American Academy of Political and Social Science* 18, America's Race Problems (July 1901), 81.

13 Bender, *American Abyss*, 124–125.

14 Baldwin, *Chicago's New Negroes*, 27–30.

15 Reed, *Not Alms but Opportunity*, 20.

16 Quoted in Lyman, *Militarism, Imperialism, and Racial Accommodation*, 120–140.

17 Quoted in Baldwin, *Chicago's New Negroes*, 27.

18 Alfred Stone, "Is Race Friction Between Blacks and Whites in the United States Growing and Inevitable?" *American Journal of Sociology* 13, no. 5 (1908), 692.

19 Gilman, "A Suggestion on the Negro Problem," 78.

20 Quoted in McKee, *Sociology and the Race Problem*, 29.

21 On the contemporaneous development of pathological models of class or class as pathology, see Mark Pittenger, *Class Unknown: Undercover Investigations of American Work and Poverty from the Progressive Era to the Present* (New York: New York University Press, 2012), 38–39.

22 O'Connor, *Poverty Knowledge*, 95.

23 Carter, *The United States of the United Races*, 113.

24 Berlin, *The Making of African America*, 154–156.

25 Giddings, *The Principles of Sociology*, 328.

26 Guterl, *The Color of Race*, 49. See Odom's theory of the "Black Ulysses" to see how this played out in the shifting labor—specifically lumber—economy of the South and how southern blacks were also seen to be incompatible with modernity. Jones, *The Tribe of Black Ulysses*.

27 Jones, *The Tribe of Black Ulysses*, 328.

28 O'Connor, *Poverty Knowledge*, 17.

29 While still a student at Fisk, Haynes arranged for DuBois to give the 1898 commencement address, which coincided with the tenth anniversary of DuBois's own graduation. Perlman, "Stirring the White Conscious," 176.

30 Haynes, *The Negro at Work in New York City*.

31 Perlman, "Stirring the White Conscious," 42.

32 Haynes, *The Negro at Work in New York City*, 44.

33 Baldwin, *Chicago's New Negroes*, 38–39.

34 Haynes, *The Negro at Work in New York City*, 44.

35 Ibid., 33.

36 For an extended critique by Haynes of Hoffman's index of the Negro death rate in southern cities, see Haynes, *The Negro at Work during the War*, 34–38, 42–44.

37 Ibid., 45–55.

38 Reed, *Not Alms but Opportunity*, 22.

39 Wayne, *Imagining Black America*, 45.

40 On the interracial dynamics of political radicalism, see Michael Kazin, *American Dreamers: How the Left Changed a Nation* (New York: Knopf, 2011).

41 Reed, *Not Alms but Opportunity*, 20.

42 Ibid. Decades before E. Franklin Frazier's *The Negro Family in Chicago*, Haynes's *The Negro at Work in New York City* was one of the best known works regarding the racial dimensions of urban ecology theory.

43 *New York Times*, January 28, 1912; Albert O. Wright, "The New Philanthropy," in *Proceedings of the National Conference of Charities and Corrections* (Boston, 1896), 4.

44 *New York Times*, January 28, 1912. In 1900, 80.9% of the black population of Manhattan was contained within twelve of the thirty-five Assembly districts— one-third of which was concentrated in only three districts: the Eleventh (10.4%), the Nineteenth (13.8%), and the Twenty-Seventh (9.2%). Haynes, *The Negro at Work in New York City*, 48–49.

45 Roediger and Esch, *Production of Difference*, 173–174.

46 See Roediger, *Working Toward Whiteness*.

47 Parris and Brooks, *Blacks in the City*, 191.

48 Reed, *Not Alms but Opportunity*, 191.

49 Ibid.

50 Emmett J. Scott, *Negro Migration during the War* (New York, 1923), 54–55.

51 Quoted in Scott, *Negro Migration during the War*, 54–55.

52 G. E. Haynes, "Cooperation with Colleges in Securing and Training Negro Social Workers for Urban Centers," *Proceedings of the National Conference of Charities and Corrections* (1916).

53 Haynes launched the Urban League Fellows Program in conjunction with the New York School of Philanthropy, which funded graduate training at the Master's level in sociology, economics, or social work at a number of universities. This provided the NUL with highly skilled employees at no cost and avoided the need

to integrate placements. The Fellows Program became a feeder program for NUL branch executives and supplied many future black social scientists such as Abram Harris and Ira Reid. Wilson, *The Segregated Scholars*, 86.

54 *The Pittsburgh Survey*, 6 vols. (Pittsburgh, PA: Russell Sage Foundation, 1909–1914); William P. Dillingham, *U.S. Immigration Commission* (Washington, DC: Government Printing Office, 1907–1911); Edward Ross, *The Old World in the New: The Significance of Past and Present Immigration to the American People* (New York: The Century Company, 1914).

55 Emily Greene Balch, "Racial Contacts and Cohesions: As Factors in the Unconstrained Fabric of a World at Peace," *Survey* (March 6, 1915), 610.

56 Wilson, *The Segregated Scholars*, 128.

57 Bender, *American Abyss*, 242–243; Wilson, *The Segregated Scholars*, 128.

58 Wilson, *The Segregated Scholars*, 128–129.

59 Quoted in ibid., 128–129. In 1907, DuBois issued a public statement rejecting erroneous suggestions that he—rather than Jackson—had prepared the Jamestown exhibit, attacking the whole affair as a "shameful and discredited enterprise."

60 Ibid.

61 George Haynes, "Negroes Move North," pt. 1, *Survey* 40 (May 4, 1918); Haynes, "Negroes Move North," pt. 2, *Survey* 41 (January 4, 1919).

62 Reed, *Not Alms but Opportunity*, 21–25.

63 Ibid.

64 On professional expertise—in the social sciences, specifically—as a form and or agent of masculine racial uplift in early twentieth-century America, see Ross, *Manning the Race*, 145–200.

65 Hall later returned to the Census Bureau, where he was the chief specialist on black population matters for a number of years. Hall had already written a number of census reports on black migration. In 1917, Hall and Jennifer were "loaned" to the DOL to do a report on blacks, which was used to justify the creation of the DNE. Wilson, *The Segregated Scholars*, 131.

66 Guzda, "Social Experiment of the Labor Department," 20.

67 Ibid.

68 Haynes, *The Negro at Work during the War*, 130, 137–138. On race and time-work discipline with a particular emphasis on the antebellum sense of the temporal, see Smith, *Mastered by the Clock*.

69 Haynes, *The Negro at Work during the War*, 137–138.

70 "Visiting Hog Island," September 1918, Records of the U.S. Shipping Board, RG 32, File Folder 12, National Archives and Records Administration–Mid-Atlantic Region, Philadelphia, PA.

71 Ibid.

72 Ibid.

73 Bruce Nelson, *Divided We Stand: American Workers and the Struggle for Black Equality* (Princeton, NJ: Princeton University Press, 2001), 9.

74 Guzda, "Social Experiment of the Labor Department," 21.

75 Haynes, *The Negro at Work during the War*, 137; *New York Times*, February 17, 1918.

76 *New York Times*, February 17, 1918.

77 Haynes, *The Negro at Work during the War*, 62.

78 Ibid.,120–130.

79 A survey of five major stockyards found that 2,990 black women (91% of those black women working in the industry) were employed in cleaning and curing offal. In a survey of sixteen tobacco plants, 5,965 women were employed as steamers (72% of the total number of employed black women). Haynes, *The Negro at Work during the War*, 125, 130.

80 On the intra-racial gender politics of racial uplift, see Mitchell, *Righteous Propagation*; Ross, *Manning the Race*.

81 Haynes, *The Negro at Work during the War*, 141.

82 Ibid., 130.

83 Ibid., 126–127.

84 Ibid.

85 Guzda, "Social Experiment of the Labor Department," 21.

86 Department of Negro Economics, Summer Memo 1918, Records of the War Production Board, RG 179, Box 21, Folder, 6, NAB II; Wilson, *The Segregated Scholars*, 131.

87 Wilson, *The Segregated Scholars*, 130–134.

88 Ibid., 132.

89 Ibid., 134.

90 Guterl, *The Color of Race*, 12.

91 Painter, *Standing at Armageddon*, 279.

92 Guzda, "Social Experiment of the Labor Department," 31–32.

93 Ibid.

94 Wilson, *The Segregated Scholars*, 132.

95 War Department Memo, April 22, 1919, Records of the War Department General and Special Staffs, RG 165.2, Box 10, Folder 2, NAB II.

96 Ibid.

97 Postwar black unemployment in the North, despite being rooted in a general slowing of the economy (labor surplus), was also linked to blacks' refusal to work at prewar rates and the demand for agricultural labor in the South. See Memo to Woods, April 19, 1919, RG 165.2, Box 10, Folder 3, NAB II.

98 War Department Memo, April 22, 1919, RG 165.2, Box 10, Folder 2, NAB II.

99 Ibid.

100 Guzda, "Social Experiment of the Labor Department," 32; Memo to Colonel Woods, Department of Labor, USES, April 19, 1919, RG 179, NAB II.

101 Ibid., 30–34.

102 Ibid., 30–34.

103 Ibid., 33.

104 Ibid., 33.

105 U.S. Congress, House, House Document 1906, 64th Congress, 2nd session; U.S. Department of Labor, *Regulations of the Department of Labor* (Washington, DC: Government Printing Office, 1915), 119–121.

106 Ovington to Secretary Wilson, Records of the National Association for the Advancement of Colored Peoples (NAACP), Box I:C80, Reel 14, Library of Congress (LOC); Eugene Kinckle Jones, "Department of Negro Economics," Records of the National Urban League (NUL), Box I:E30, LOC.

107 Guzda, "Social Experiment of the Labor Department," 35.

108 U.S. Congress, Civil Sundry Bill, Press Clippings, Records of the NAACP, Box C319, LOC.

109 Roediger, *Working Toward Whiteness*, 7.

110 Jane Schiber and Harry Schiber, "The Wilson Administration," 449; Wilson, *The Segregated Scholars*, 131.

111 Haynes, *The Negro at Work during the War*, 7.

112 Ibid., 7–9.

113 Ibid., 7–9.

114 Wilson, *The Segregated Scholars*, 131.

115 Charles Johnson, "Black Workers and the City," *Survey Graphic* 6, no. 6 (March 1925); Abram Harris, "The Economic Foundations of American Race Division," *Social Forces* 5, no. 3 (March 1927), 468–478; Harris, *"The Negro Population in Minneapolis: A Study of Race Relations,"* National Urban League (1926); Harris, *The Negro as Capitalist* (1936).

116 For prior sociological examinations of black proletarianization, see Charles Harris Wesley, *Negro Labor in the United States, 1850–1925: A Study in American Economic History* (New York: Vanguard Press, 1927).

117 Roediger and Esch, *The Production of Difference*, 161–162.

118 Montgomery, *Workers' Control in America*, 122.

119 Bay, *The White Image in the Black Mind*, 202–203.

120 Lipsitz, *The Possessive Investment in Whiteness*, viii.

121 Roediger, *The Wages of Whiteness*, 11–13.

122 Holt, *Children of Fire*, 4.

123 For insights into the relation of race to industry, see Michael Adas, *Machines as the Measure of Men: Science, Technology, and Ideologies of Western Dominance* (Cornell University Press, 1990); Gilman, *Difference and Pathology*; Bender, *American Abyss*, 13–15. For a brilliant cultural/artistic perspective of blacks' relation to and within American modernity—inhabiting and shaping modernity in sonic terms, see Dinerstein, *Swinging the Machine*.

124 In arguing that the ideal working man (and manager) "was made" and not born, Taylor seemed to be echoing an environmentalist view of social difference—a view that was potentially sympathetic to groups such as blacks who were generally seen as slaves to their nature. Taylor, *The Principles of Scientific Management*, iv. For work on Taylorism and race, see Frader, "From Muscles to Nerves"; Lyman, *Militarism, Imperialism and Racial Accommodation*, 120–140.

125 Brown, "The Negro Migrant."

126 Miles, *Capitalism and Unfree Labor*; DuBois, *Darkwater*, 34–35.

127 On the structure of the prewar/wartime state, see Hawley, *The Great War*; Alan Dawley, *Struggles for Justice: Social Responsibility and the Liberal State* (Cambridge: Belknap Press of Harvard University Press, 1991).

CHAPTER 3. MEASURING MEN FOR THE WORK OF WAR

1 H. Berry Memoir, *A Day in the Army at Tuskegee Institute*, Berry Family Letters, Schomburg Center for Research in Black Culture, New York Public Library (NYPL).

2 A process was often referred to as the "inspection effect" of wartime economies. Bourke, *Dismembering the Male*, 171–179.

3 Bristow, *Making Men Moral*, 3.

4 Roediger, *How Race Survived U.S. History*, 160–161.

5 Bourne, "The State," (1918); *The Untimely Papers* (1919). Bourne's critique of the wartime state as a mechanism of corporeal control is further complicated and perhaps informed by his experiences as an American living with a physical disability. Bourne's face was disfigured at birth by the misuse of forceps and his spine was deformed by a childhood bout of tuberculosis. For Bourne's reflections on disability as both an identity and category of analysis, see Randolph Bourne, "The Handicapped," *Youth and Life*, 1913.

6 Irvin Cobb, quoted in Slotkin, *Lost Battalions*, 153–154; Gerstle, *American Crucible*, 84–85.

7 Ales Hrdlicka, quoted in Frederick L. Hoffman, *Army Anthropometry and Medical Rejection Statistics* (Prudential Press, 1918), 14–16.

8 Franz Boas, "Changes in Bodily Form of the Descendants of Immigrants," *American Anthropologist* 14, no. 3 (July–September 1912), 530–562.

9 Hoffman, *Army Anthropometry*, 16, 54. Davenport and Yerkes had corresponded at the beginning of the war regarding the linkages between heredity and mental and the physiological development. In a letter to Yerkes, Davenport remarked: "I am now rushing a book on the subject of naval officers with reference to their juvenile and family history. It is interesting to note that Admiral T Mahan was the one man who has expressed very clearly the idea that the effectiveness of a man in his occupation depends upon his hereditary traits, together with the opportunities that they have for development and exercise." Davenport to Yerkes, May 16, 1917, Charles B. Davenport Papers, Series 1, Box 97, Folder 5, American Philosophical Society (APS).

10 Ireland, Davenport, and Love, "Part One: Army Anthropology," 45.

11 Bender, *American Abyss*, 112.

12 Bederman, *Manliness and Civilization*; Boris and Baron, "The Body as a Useful Category," 24, 25. For nineteenth-century antecedents of theories of racial embodiment, see Sappol, *A Traffic of Dead Bodies*; Roediger, *The Wages of Whiteness*, 3–5.

13 Baron, "Masculinity, the Embodied Male Worker," 143–160; Jones, *American Work*, 336; Boris and Baron, "The Body as a Useful Category," 34. For further analysis on the malleability and contingent nature of racial ideology—along corporeal lines or otherwise, see Baker, *Anthropology and the Racial Politics*; Roediger, *Colored White*; Gilroy, *Against Race*, 11–15.

14 Holmes to Hrdlicka, Report of the Committee on Anthropology, *Proceedings of the National Academy of Sciences* (1918).

15 For insight into how DuBois engaged with this scholarship, used it to bind his scientific and literary work, and reconfigured it as a means of uplift, see Farland, "W. E. B. DuBois, Anthropometric Science."

16 Bruinius, *Better for All the World*, 162; Kelves, *In the Name of Eugenics*, 44–46.

17 Kelves, *In the Name of Eugenics*, 46–47.

18 Dawley, *Changing the World*, 194.

19 Charles Johnson Post, "The Army as a Social Service," *Survey* (May 20, 1916), 201.

20 Roger Horowitz, "'It is the Working Class Who Fight All the Battles': Military Service, Patriotism and the Study of American Workers," in Halpern and Morris, *American Exceptionalism?*, 76–100; Shenk, "*Work or Fight!*," 6–7.

21 Minutes of a Meeting of the Committee of Anthropologists at the NRC, Washington, DC, November 15, 1918, Davenport Papers, Series 1, Box 81, Folder 18, APS (hereafter cited as Minutes, Committee of Anthropologists at the NRC).

22 Though historians have debated the actual extent to which Taylorist models were adopted by business, many still contend that he was "the most influential management *theorist* of his time." McCartin, *Labor's Great War*, 3, 50.

23 Minutes, Committee of Anthropologists at the NRC.

24 Braun, "Spirometry, Measurement, and Race."

25 Gould, *Investigations in the Military*, 477–497. Historian Margaret Humphreys disputes this assertion, claiming that Gould himself did not draw this conclusion but merely documented the measurement. Humphreys, *Intensely Human*, 152.

26 Sanford Hunt, "The Negro as Soldier," *Anthropological Review* 7, no. 24 (January 1869), 40–54.

27 On the era's corpus of black extinction literature, see Lawrie, "'Mortality as the Life Story of a People,'" *Canadian Review of American Studies* (2013) 375.

28 Vernon Williams Jr., "What is Race? Franz Boas Reconsidered," in Campbell, *Race, Nation, and Empire*, 40–64. The intersections of race, imperialism, and anthropology revealed themselves in intriguing ways. In 1904, Boas contacted Booker T. Washington on behalf of J. E. Aggrey, a "full blooded Negro" and a student at Livingstone College in Salisbury, North Carolina who wished to study anthropology at Columbia University. While Boas acknowledged that he was hesitant to "advise the young man to take up this work for fear that it would be difficult for him to find a place" following his studies, he was hopeful that Aggrey could find work in the Colonial services of a European African colony. Washington, however, confirmed Boas's initial doubts, arguing that Aggrey's proposed course of study would "be of little value to him" given his

race. Boas to Washington, November, 30, 1904, Franz Boas Papers, APS, https://diglib.amphilsoc.org/boas-request?key=text:130178; Washington to Boas, December 9, 1904, Boas Papers, APS, https://diglib.amphilsoc.org/ boas-request?key=text:130180; DuBois to Boas, October 11, 1905, Boas Papers, APS, http://diglib.amphilsoc.org/islandora/object/text:39375.

29 Haller, *Outcasts from Evolution*, 34.

30 Grant, *The Passing of the Great Race*, 74.

31 Benjamin Kidd, quoted in Anderson, *Colonial Pathologies*, 41.

32 On the climatic and environmental dimensions of race and imperial labor, see Bender, *American Abyss*, 40–68.

33 William Washburn, "The Relation Between Climate and Health with special reference to American Occupation of the Philippine Islands," *American Journal of the Medical Sciences* (1905), 515.

34 On the intersections of race and labor in the broader logic of American imperialism, see Bender and Lipman, *Making the Empire Work*.

35 N. S. Shaler, "The Future of the Negro in the Southern States," *Popular Science Monthly* 57 (June 1900), 150.

36 Joseph O. Baylen and John Hammond Moore, "Senator John Tyler Morgan and Negro Colonization in the Philippines, 1901 to 1902," *Phylon* 29, no. 1 (1968), 65–75.

37 Bender, *American Abyss*, 88; Anderson, *Colonial Pathologies*, 102–103. On blacks'—specifically black troops'—fractious relationship to U.S. imperialism, see Murphy, *Shadowing the White Man's Burden*.

38 Bender, *American Abyss*, 129.

39 Minutes, Committee of Anthropologists at the NRC. For European perceptions of conscription's dysgenic consequences, see Pick, *Faces of Degeneration*.

40 Gilman, "A Suggestion on the Negro Problem," 78–85.

41 Bourke, *Dismembering the Male*, 175; Hoffman, *Army Anthropometry*, 41–44.

42 Hoffman, *Army Anthropometry*, 42–44.

43 Ibid., 49; Bender, *American Abyss*, 66–67.

44 *Scientific America*, June 9, 1917.

45 Robert De C. Ward, "Some Aspects of Immigration to the United States in Relation to the Future of the American Race," *Eugenics Review* 7 (April 1915–January 1916), 263–282. For further insights into the dysgenic nature of war, see Jordan, *War and Waste*.

46 T. J. Downing, "A Possible Factor of Degeneracy," *New York Medical Journal* 108 (July 20, 1917), 103–105.

47 Hoffman, *Army Anthropometry*, 16.

48 Kennedy, *Over Here*, 144–154.

49 Those responsible for raising a new national army, such as Secretary of War Newton Baker and Provost Marshal General Enoch Crowder, head of the Selective Service, were keen students of the Civil War and were well aware of the fact that the draft had yielded only 6% of the Union manpower and resulted in

widespread civil unrest such as the bloody 1863 draft riots in New York City. Kennedy, *Over Here*, 68; Painter, *Standing at Armageddon*, 331.

50 Hrdlicka to Davenport, February 4, 1918, Davenport Papers, Series 1, Box 53, Folder 2, APS; Kennedy, *Over Here*, 149.

51 Hrdlicka to Davenport, February 4, 1918, Davenport Papers, Series 1, Box 53, Folder 2, APS.

52 Davenport and Love, *Physical Examination of the First Million Draft Recruits*, 25.

53 Hoffman, *Army Anthropometry*, 114; Office of the Provost Marshal General, *Standards of Physical Examination*, Form 75, 4.

54 DuBois, quoted in Williams, *Torchbearers of Democracy*, 77.

55 Ibid., 25, 75.

56 Slotkin, *Lost Battalions*, 239; Shenk, "*Work or Fight!*," 4–10.

57 For insight into this tradition and how it conflated with blacks' understandings of uplift ideology and imperialism, see Gaines, *Uplifting the Race*, 27; Mitchell, "'The Black Man's Burden,'" 77–99.

58 "Stevedores' Career a Round of Harmony," June 7, 1918, *Stars and Stripes*.

59 Franklin, *From Slavery to Freedom*, 455, 462–463; Buckley, *American Patriots*, 165.

60 Ireland, Davenport, and Love, "Part One: Army Anthropology."

61 Henry Fairfield Osborne, "The Fighting Ability of Different Races," *Journal of Heredity* 10, no. 1 (1919), 29–31.

62 Williams, *Torchbearers of Democracy*, 108, 110–112.

63 For the racial dynamics of Progressive Era penal labor, see Liechtenstein, *Twice the Work of Free Labor*.

64 Davenport and Love, *Army Anthropology*, 35. In one instance, the Surgeon General's Office subsection of anthropology (established in July 1918 under Davenport's supervision) was called upon to intervene in a case of disputed classification of a recruit who claimed not to have colored blood. Unfortunately, the archival record does not reveal the outcome of this dispute. What is significant is the way in which some officials believed that anthropometric evaluation of said recruit could provide a reliable index of his racial typology. This provides further evidence of anthropometry's role in linking race and color—not to traits and tendencies, but to the body. Davenport to Grant, January 12, 1918, Davenport Papers, Series 1, Box 42, Folder 21, APS.

65 Extract from letter of Professor L. Manouvrier, Laboratoire d'Anthropologie, École Pratique des Hautes Études, Paris, December 25, 1917 to Dr. A. Hrdlicka, March, 12, 1918, National Research Council (NRC) Collection, APS.

66 Berry, *A Day in the Army at Tuskegee Institute*; Davenport and Love, *Army Anthropology*, 34.

67 James Mennell, "African Americans and the Selective Service Act of 1917," *Journal of Negro History* (1982); Painter, *Standing at Armageddon*, 332.

68 Kennedy, *Over Here*, 162; Barbeau and Henri, *The Unknown Soldiers*, 80; Shenk, "*Work or Fight!*," 45.

69 Hoffman to Davenport, August 5, 1918, Davenport Papers, Series 1, Box, 52, Folder 2, APS.

70 Ibid.

71 Minutes, Committee of Anthropologists at the NRC.

72 Hoffman, *Army Anthropometry*, 15.

73 Hale to Hoffman, November 22, 1917, Re: U.S. Army Anthropometric Work, Davenport Papers, Series 1, Box 44, Folder 1, APS.

74 Hoffman, *Army Anthropometry*, 15; Davenport to Grant, December, 31, 1917, Davenport Papers, Series 1, Box 42, Folder 5, APS.

75 See Williams Jr., "What is Race? Franz Boas Reconsidered."

76 Office of the Surgeon General U.S. Army, "Malingering in U.S. Troops, Home Forces 1917," *Military Surgeon* 42, no. 3 (March 1918), 21–25.

77 Ibid.; Taylor, *The Principles of Scientific Management*, 6–7.

78 Office of the Provost Marshal General, *Standards of Physical Examination*, Form 75.

79 Slotkin, *Lost Battalions*, 63.

80 Office of the Provost Marshal General, *Standards of Physical Examination*, Form 75, 7.

81 Peter Way, "'Black Service . . . White Money': The Peculiar Institution of Military Labor in the British Army during the Seven Years' War," in Leon Fink, ed., *Workers Across the Americas: The Transnational Turn in Labor History* (New York: Oxford University Press, 2010).

82 On the temporal dynamics of blacks transition from bondage to contract labor, see Stanley, *Bondage to Contract*.

83 Kennedy, *Over Here*, 150.

84 Office of the Provost Marshal General, *Manual of Instructions for Medical Advisory Boards*, Form 64, 2.

85 Ibid.; Office of the Provost Marshal General, *Manual of Instructions for Medical Advisory Boards*, Form 65, 3; Davenport and Love, *Physical Examination of First Million Draft Recruits*, 23.

86 Office of the Provost Marshal General, *Manual of Instructions for Medical Advisory Boards*, Form 65, 3.

87 K. Walter Hickel, "Medicine, Bureaucracy, and Social Welfare," in Longmore and Umansky, *The New Disability History*, 256.

88 Davenport and Love, *Physical Examination of the First Million Draft Recruits*, 22. F. Hoffman claimed that confusing venereal diseases with physical deficiencies of the body was "as serious an error as to confuse organic defects of the lungs with defects of lung capacity." Hoffman, *Army Anthropometry*, 13.

89 Davenport and Love, *Army Anthropology*, 22.

90 Hrdlicka to Hale, February, 4, 1918; Hrdlicka to Hale, July 1918, Davenport Papers, Series 1, Box 53, Folder 3, APS.

91 Ibid.

92 Davenport and Love, *Army Anthropology*, 35.

93 Bender, *American Abyss*, 240.

94 Spiro, *Defending the Master Race*, 311; Davenport to Grant, January 12, 1918, Davenport Papers, Series 1, Box 42, Folder 6, APS. The Surgeon General's Office could be forgiven for their rectitude given Davenport's more extreme and baffling statements—such as his stubborn belief that a dominant gene for *thalassophilia* (love of the sea) predisposed its carriers to careers as naval captains. Witkowski and Inglis, *Davenport's Dream*, 7.

95 Grant to Davenport, July 6, 1914, Davenport Papers, Series 1, Box 42, Folder 3, APS.

96 While eugenics had already started to receive criticism within the scientific community as early as the 1910s—thanks to the work of Boas and his students at Columbia—the public, or popular, incarnation of eugenics flourished during the 1920s and 30s. For analysis of the lag between scientific and popular understandings of eugenics, see Currell and Cogdell, *Popular Eugenics*.

97 Davenport and Love, *Army Anthropology*, 50.

98 Ibid., 50.

99 Ibid., 17.

100 Ibid., 17.

101 Grant to William Gregory, American Museum of Natural History, April 18, 1918, Eugenics Records Office Records, APS.

102 Guterl, *The Color of Race*, 30–31.

103 Guterl, *The Color of Race*, 30–32; Spiro, *Defending the Master Race*, 304; Kelves, *In the Name of Eugenics*, 75.

104 Spiro, *Defending the Master Race*, 314; Baker, *From Savage to Negro*: 93.

105 "Race in Relation to Disease Civilian Records," May 1918, Hoffman Papers, Box 13, Item 57, Butler Library Rare Book and Manuscripts, Columbia University.

106 Ibid.

107 Ibid.

108 Minutes of the U.S. Shipping Board, Spring 1918, RG 32, File 25, Folder 2, National Archives and Record Administration–Mid-Atlantic Region, Philadelphia, PA (hereafter cited as Minutes, U.S. Shipping Board); Martin, *The Saga of Hog's Island*, 34–35.

109 Quoted in Martin, *The Saga of Hog's Island*, 34–35.

110 *New York Times*, February 17, 1918.

111 Minutes, U.S. Shipping Board, Spring 1918.

112 Ibid.

113 Charles Goring, *The English Convict* (1913); Kelves, *In the Name of Eugenics*, 70–71; Piers Beirne and James W. Messerschmidt, *Criminology*, 4th ed. (Los Angeles, CA: Roxbury, 2005), 82.

114 Hoffman to Pearce, July 15, 1918, Davenport Papers, Series 1, Box 52, Folder 3, APS.

115 Hoffman to Davenport, August 5, 1918, Davenport Papers, Series 1, Box 52, Folder 3, APS.

116 Ibid.

117 Ibid.

118 Ibid.

119 Minutes, U.S. Shipping Board, Spring 1918.

120 "U.S. to Act to Oust Ship Work Slackers," *New York Tribune,* September 22, 1918, 9.

121 Ibid.

122 To Franz Boas, November 20, 1919, National Research Council–Division of Anthropology and Psychology, Boas Papers, APS, https://diglib.amphilsoc.org/boas-request?key=text:91232.

123 Hoffman to Matthew Well, Committee on Labor, Council of National Defense, September, 11, 1918, Davenport Papers, Series 1, Box 52, Folder 3, APS.

124 Hoffman to Davenport, September 4, 1918, Davenport Papers, Series 1, Box 52, Folder 3, APS.

125 Hoffman to Woll, Assistant to Gompers, September, 11, 1918; Hoffman to Davenport, April 15, 1919, Davenport Papers, Series 1, Box 52, Folder 3, APS; Davenport to Boas, May, 2, 1919, Davenport Papers, Series 1, Box 5, Folder 2, APS.

126 Hooton to Baker, November 22, 1918, Davenport Papers, Series 1, Box 52, Folder 2, APS.

127 Ibid.; Davenport and Love, *Army Anthropology,* 56.

128 Spiro, *Defending the Master Race,* 312.

129 Ibid.

130 Davenport and Love, *Army Anthropology,* 47, 54.

131 Davenport to Merriam, Re: Memo for the Adjunct General of the Army, June 9, 1919, Davenport Papers, Series 1, Box 101, Folder 3, APS.

132 Davenport and Love, *Army Anthropology,* 59–61.

133 Ibid., 59–61.

134 Ibid., 34, 40.

135 This hypothesis drew in part on Boas's previous prewar work on shifting immigrant physiognomies, *Changes in Bodily Form of Descendants of Immigrants* (1910). In response to critics, Boas would go on to publish over 500 pages of raw data from this study entitled *Materials for the Study of Inheritance in Man* (1928). Davenport and Love, *Army Anthropology,* 59.

136 Davenport to the Surgeon General of the Army, July 31, 1922; June 25, 1919, Davenport Papers, Series 1, Box. 12, Folder 2, APS.

137 Davenport and Love, *Army Anthropology,* 35–39.

138 Ibid.

139 *Literary Digest,* June 14, 1919, 23.

140 Roediger and Esch, *The Production of Difference,* 182.

141 Gelber, "A 'Hard-Boiled Order,'" 167–168.

142 Yerkes, *Psychological Examining in the U.S. Army*; Gould, *The Mismeasure of Man,* 227; Kelves, *In the Name of Eugenics,* 80–83.

143 *Literary Digest,* June 14, 1919, 23.

144 Brad Campbell, "The Making of 'American': Race and Nation in Neurasthenic Discourse," *History of Psychiatry* 18, no. 2 (2007), 162.

145 Ibid., 23.

146 Minutes, Committee of Anthropologists at the NRC.

147 Charles Davenport, "The Measurement of Men," (lecture, American Association for the Advancement of Science, Philadelphia, PA, January, 14, 1927), Davenport Papers, Series 1, Box 27, APS.

148 Gould, *The Mismeasure of Man*, 227; Minutes.

149 Minutes, Committee of Anthropologists at the NRC.

150 Charles Davenport, "Aims and Methods in Anthropometry," (lecture, Institut International d'Anthropologie, September 23, 1927), Davenport Papers, Series 1, Box 24, APS; Stiker, *A History of Disability*, 124.

CHAPTER 4. SALVAGING THE NEGRO

1 Hickel, "Medicine, Bureaucracy, and Social Welfare," 256–258.

2 Harry Mock, "Reclamation of the Disabled from the Industrial Army," *Annals of the American Academy of Political and Social Science* 80, Rehabilitation of the Wounded (November 1918).

3 Despite the public's fascination with the amputee, fewer than five thousand men in need of prosthetics returned to the United States. Lansing, "Salvaging the Manpower," 2. Prior to the war, the U.S. was one of the world's largest producers of artificial limbs. Linker, *War's Waste*, 98–99.

4 See Stoddard, *The Rising Tide of Color*; Stoddard, *The Revolt Against Civilization: The Menace of the Under Man* (New York: Scribner, 1922), 120–125.

5 Gerber, *Disabled Veterans in History*, 4; "Caring for the Soldiers' Health: Reducing the Loss from Sickness and Wounds, Businesslike Humanity—Burying 140 Men an Hour," *World's Work* 28, no. 6 (October 1914), 119.

6 Gerber, *Disabled Veterans in History*, 4.

7 Deborah Cohen, "Will to Work: Disabled Veterans in Britain and Germany after the First World War," in Gerber, *Disabled Veterans in History*, 297; Keane, *Doughboys, the Great War, and the Remaking of America*, 6, 163; Federal Board of Vocational Education, "What Every Disabled Soldier and Sailor Should Know"; Gelber, "A 'Hard-Boiled Order,'" 167.

8 For further insight into anthropological notions of the Negro mind and body divide in modern American racial thought, see Baker, *From Savage to Negro*; Scott, *Contempt and Pity*.

9 Gerber, *Disabled Veterans in History*, 4; Bourke, *Dismembering the Male*. For insights into rehabilitation as a mechanism of colonial and transnational racial labor control, see West, "Divine Fragments," 22–25; Frader, "From Muscles to Nerves," 123–147; Jennifer Keene, "Protest and Disability: A New Look at African American Soldiers During the First World War" in *Warfare and Belligerence: Perspectives in First World War Studies*, ed. Pierre Purseigle (Boston: Brill, 2005), 216–241.

10 Degler, *In Search of Human Nature*, viii–ix, 75–78; Banta, *Taylored Lives*, 4.

11 Beth Linker, "Feet For Fighting: Locating Disability and Social Medicine in First World War America," *Social History of Medicine* 20, no. 1 (April 2007), 91–109; Bender, *American Abyss*, 243–245; Price, "Lives and Limbs," 5.

12 Elizabeth Upham, "Selective Placement of the Disabled," *Vocational Summary* 2, no. 2 (June 1919), 35.

13 Longmore, *Why I Burned My Book*, 2.

14 Boris and Baron, "The Body as a Useful Category," 23; Price, "Lives and Limbs," 5; Keough, "War-Disabled Negroes in Training," *Vocational Summary* (Spring 1923), 33.

15 By 1930, half a million blacks had left the region of their birth. At mid-century, 96% of black northerners and 90% of black westerners lived in urban areas. Ira Berlin, *The Making of African America*, 154.

16 Ibid., 4–10.

17 Franklin, *From Slavery to Freedom*, 455, 462–463; Buckley, *American Patriots*, 165; Keene, "Protest and Disability," 218–219.

18 Williams, *Torchbearers of Democracy*, 108–112.

19 Kennedy, *Over Here*, 199; Barbeau and Henri, *The Unknown Soldiers*, 191; Budreau, *Bodies of War*, 54–55.

20 Bay, *The White Image in the Black Mind*, 204; Slotkin, *Lost Battalions*; "Returned Negro Soldiers," *Survey* (May 3, 1919), 207; James Sanford to Dr. J. R. A. Crossland, "Colored Physicians to Attend Colored Vocational Men," November 10, 1921, Records of the Department of Veterans Affairs, RG 15, Entry 50, Box 2, General Correspondence File Folder, National Archives Building, Washington, DC (NAB); W. E. B. DuBois, "Returning Soldiers," *The Crisis*, (May 1919).

21 Williams, *Torchbearers of Democracy*, 224–225.

22 Kennedy, *Over Here*, 214–225; "Returned Negro Soldiers," 207.

23 See Paul R. D. Lawrie, "Salvaging the Negro: Race, Rehabilitation, and the Body Politic in World War One America," in Burch and Rembis, *Disability Histories*, 321–344.

24 Bourke, *Dismembering the Male*, 171–179, refers to this process as the "inspection effect" of wartime economies.

25 One of the first progressives to cite the racially dysgenic effects of war was David Starr Jordan, the founding president of Stanford University. See Jordan, *War and Waste*; Bender, *American Abyss*, 233–234; Gerber, *Disabled Veterans in History*, 8; Hussie, "Forestry and Tree Culture," 725–726.

26 Croly, quoted in Harry Mock, "Human Conservation and Reclamation," *American Journal of Care for Cripples* 6, no. 1, 5. For analysis of the shift from social forms of "manhood" to somatic understandings of "masculinity and its martial character" in progressive America, see Kasson, *Houdini, Tarzan, and the Perfect Man*; Bederman, *Manliness and Civilization*.

27 Keough, "War-Disabled Negroes in Training," 33.

28 Mock, "Reclamation of the Disabled," 30.

29 Office of the Surgeon General, *The Medical Department of the United States Army*, 2, 28; Paul Kellogg, "A Canadian City in War Time: The Battle-Ground for Wounded Men," *Survey* (April 7, 1917), 1–2; Thomas Gregory, "Restoring Crippled Soldiers to a Useful Life," *World's Work* (1918), 427–432; John Todd, "The Meaning of Rehabilitation," *Annals of the American Academy of Political and Social Science* 80, Rehabilitation of the Wounded (November 1918), 1.

30 Douglas McMutrie, "Your Duty to the War Cripple," *American Journal of Care for Cripples* 7, 82; "The High Road to Self-Support," *Carry On* 1, no. 1, 4–9; George Price, "Rehabilitation Problems," *Survey* (March 29, 1919), 921–922; W. S. Bainbridge, "Social Responsibilities in the Rehabilitation of Disabled Soldiers and Sailors," *American Journal of Care for Cripples* 7, no. 2, 126–132.

31 Holt, *The Federal Board of Vocational Education*, 45.

32 Ibid.

33 Linker, *War's Waste*, 33.

34 The FBVE took up the administration of this work until it was subsumed by the Veterans Bureau in 1921. Thurber, *Preliminary Checklist*, 6.

35 Rupert Hughes, "The Lucky Handicap," *Carry On* 1 (September 1918), 11–12; James Munroe, "The War's Crippled: How They May Be Made Assets Both to Themselves and to Society," *Survey* (May 18, 1918), 28; Gelber, "A 'Hard-Boiled Order,'" 167. Black Civil War veterans were especially vulnerable to charges of "dependency" and were subsequently denied pensions at nearly double the rate of their white peers. For insight into late nineteenth-century discourses of race, health, and citizenship, see Downs, *Sick from Freedom*, 156–158.

36 McMutrie, "Your Duty to the War Cripple," 82.

37 Bainbridge, "Social Responsibilities," 126–132.

38 Quoted in Hughes, "The Lucky Handicap," 11–12.

39 Gelber, "A 'Hard-Boiled Order,'" 167.

40 Munroe, "The War's Crippled," 28; ibid., 179–183; *Vocational Summary* 1, no. 3 (July 1918), 1; Mock, "Reclamation of the Disabled," 8, 28; Grace Harper, "Re-Education from the Point of View of the Disabled Soldier," *American Journal of Care for Cripples* 7, no. 2 (1918), 85.

41 Munroe, "The War's Crippled," 28.

42 "Conserve Motherhood and Childhood-Make Skilled Workers of Both War and Industrial Cripples," *Labor Laws in Wartime Special Bulletin*, The American Association for Labor Legislation, no. 4, May 1918, RG 15, Folder Entry 55, Folder 4, NAB.

43 Francis Patterson, "Industrial Training for the Wounded," *Annals of the American Academy of Political and Social Science* 80, Rehabilitation of the Wounded (November 1918), 41; E. H. Fish, "Human Engineering," *Journal of Applied Psychology* (Spring 1922), 161; Lipsitz, *The Possessive Investment in Whiteness*, 74–76.

44 Callie Hull, "Reopening Industry's Doors to the Returning Negro Soldiers," *Vocational Summary* 2, no. 5 (September 1919), 88.

45 Linker, *War's Waste*, 61.

46 Bourke, *Dismembering the Male*, 171–179.

47 Mock, "Reclamation of the Disabled."

48 Ibid.

49 Thurber, *Preliminary Checklist*, 6–7.

50 Ibid.

51 Hickel, "Medicine, Bureaucracy, and Social Welfare," 239.

52 Ibid., 235–245.

53 Bristow, *Making Men Moral*, 176–177; "Returned Negro Soldiers," 207; "Negro Soldiers and Labor," *The Crisis* (May 1919); Sergeant Greenleaf B. Johnson, "The Negro's Part in the War in Democracy," *Washington Bee*, January 18, 1919.

54 Hickel, "Medicine, Bureaucracy, and Social Welfare," 240.

55 Quoted in Ana Carden-Coyne, "Ungrateful Bodies: Rehabilitation, Resistance and Disabled American Veterans of the First World War," *European Review of History* 14, no. 4 (December 2007), 548.

56 "From His Neck Up: A Man May Be Worth $100,000 a Year," *Carry On* 1, no. 1 (June 1918), 23.

57 Ibid. For early twentieth-century theories of labor power as a metric of social efficiency, see Rabinbach, *The Human Motor*, 242–244; Luther Gulick, "The Effect of Fatigue on Character," *World's Work* (August 1907); Alice Hamilton, "Fatigue, Efficiency, and Insurance Discussed by the American Public Health Association," *Survey* 37, no. 6 (November 1916), 135–138; Mock, "Reclamation of the Disabled," 30.

58 Sociologist Robert Park famously characterized the Negro as the "lady of the races" due to blacks' perceived love of finery and profound lack of vitality. Quoted in Baldwin, *Chicago's New Negroes*, 26–27.

59 Hickel, "Medicine, Bureaucracy, and Social Warfare," 257.

60 Ibid.

61 *The Crisis* (June 1920), 193; "Returned Negro Soldiers"; Sanford to Crossland, "Colored Physicians to Attend Colored Vocational Men"; DuBois, "Returning Soldiers"; Bristow, *Making Men Moral*, 176–177; "Negro Soldiers and Labor"; Johnson, "The Negro's Part in the War in Democracy."

62 Hickel, "Medicine, Bureaucracy, and Social Warfare," 240–245; Kite, *Cotton Fields No More*, 98.

63 Keough, "War-Disabled Negroes in Training," 93.

64 Ibid.

65 Frank Billings, "Reconstruction Has Begun," *Carry On* 1, no. 2 (1918), 24.

66 Ibid.

67 *The Crisis* (June 1920), 193; Keough, "War-Disabled Negroes in Training," 93.

68 Guterl, *The Color of Race*, 12–13.

69 From 1918 to 1921, the fourteen districts were: (1) Maine, New Hampshire, Vermont, Massachusetts, Rhode Island; (2) Connecticut, New York, New Jersey; (3) Pennsylvania, Delaware; (4) D.C., Maryland, Virginia, West Virginia; (5)

North Carolina, South Carolina, Georgia, Florida, and Tennessee; (6) Alabama, Mississippi, Louisiana; (7) Ohio, Indiana, Kentucky; (8) Michigan, Illinois, Wisconsin; (9) Iowa, Nebraska, Kansas, Missouri; (10) Minnesota, Montana, North Dakota, South Dakota; (11) Wyoming, Colorado, New Mexico, Utah; (12) California, Nevada, Arizona; (13) Idaho, Oregon, Washington; and (14) Arkansas, Oklahoma, Texas. In 1921, the district offices combined with Public Health Service Offices, subsumed under the Veterans Bureau.

70 "Biography of Crossland," RG 15, Entry 50, Box 1, Misc. Correspondence File 1920–1925, NAB.

71 Ibid.

72 Todd, "The Meaning of Rehabilitation," 8.

73 Ibid.

74 Crossland to Acting Assistant Director, Rehabilitation Division, U.S. Veterans Bureau, April 3, 1923, RG 15, Entry 50, Box 1, NAB.

75 Hull, "Reopening Industry's Doors," 88–90.

76 H. O. Sargent, "Vocational Agricultural Education for Negroes in the Southern Region," *Vocational Summary* 1, no. 7 (November 1918), 12.

77 Ibid.

78 Ibid.

79 Chas. M. Griffith to Director, U.S. Veterans Bureau, April 29, 1922, RG 15, Entry 55, Folder 2, NAB.

80 Quoted in Ernest Luce to I. Fisher, "Disabled Negro Soldiers in Training," Veterans Administration, October 1920, RG 15, Entry 20, Box 1, Office Files on Supervisor of Advisement and Training, NAB.

81 J. R. Crossland to Smith, U.S. Veterans Bureau–Division of Vocational Rehabilitation, January 3, 1922, RG 15, Entry 50, Box 2, General Correspondence File Folder, NAB.

82 Wiley Hill to J. R. Crossland, January 23, 1922, RG 15, Entry 50, Box 4, NAB.

83 "War-Disabled Negroes in Training," 33; Gamble, *Making a Place for Ourselves*, 73–74.

84 Linker, *War's Waste*, 87–88.

85 Ibid.

86 Gamble, *Making a Place for Ourselves*, 73–74; *The Crisis* (Spring 1920).

87 Carden-Coyne, "Ungrateful Bodies," 547, notes that the idea of placing grown men in nurseries indicated how military medical authorities upheld ambiguous attitudes towards disabled soldiers. However, this ambiguity faded somewhat in regards to African American veterans, given prevailing social scientific and anthropological notions of blacks as a backward, childlike race in need of constant supervision.

88 Gamble, *Making a Place for Ourselves*, 73.

89 Ibid.

90 "War-Disabled Negroes in Training," 33.

91 Quoted in William J. Schieffelin, "The Most Unforgettable Character I've Met: Robert Russa Moton," *Reader's Digest*, November 1950.

92 Schneider, *We Return Fighting*, 236.

93 For an analysis of the rise of interwar forms of racial nationalism and KKK infiltration of 1920s government, see Gerstle, *American Crucible*, 120–126.

94 Schneider, *We Return Fighting*, 235–237; Clifton Dummett and Eugene Dibble, "Historical Notes on the Tuskegee Veterans Hospitals," *Journal of the National Medical Association* 54, no. 2 (March 1962), 134–135.

95 Dummett and Dibble, "Historical Notes," 135; Thomas Ruth, Director War Service, to Directors of War Service, "Service Connection on Tuberculosis and Neuropsychotic Diseases in Relation to Vocational Training," November 2, 1922, Records of the American Red Cross, 1917–1934, RG 200, Box 600, NAB II; Evelyn Phelps to Helen Ryan, "Monthly Report of the American Red Cross, Tuskegee, Alabama," June–August 1923, RG 200, NAB II.

96 Office of the Surgeon General, *The Medical Department of the United States Army*, 95.

97 Evelyn Phelps to Helen Ryan, "Monthly Report of the American Red Cross, Tuskegee, Alabama," June 1923, RG 200, NAB II; Evelyn Phelps to Helen Ryan, "Monthly Report of the American Red Cross, Tuskegee, Alabama," August 31, 1923, RG 200, NAB II.

98 E. M. Murray to Helen Ryan, The American Red Cross, U.S. Veterans Hospital No. 91, Tuskegee, Alabama, May 1, 1924, RG 200, Entry 55, Box 21, Folder 6, NAB II.

99 Phelps to Ryan, "Monthly Report of the American Red Cross," June 1923.

100 Phelps to Ryan, "Monthly Report of the American Red Cross," June–August 1923.

101 Ibid.

102 On early twentieth-century ideologies of black masculinity and racial uplift, see Summers, *Manliness and its Discontents*.

103 Phelps to Ryan, June 1924; E. M. Murray to Pauline Radford, The American Red Cross, U.S. Veterans Hospital No. 91, January 1925, RG 200, NAB II.

104 Murray to Ryan, May 1, 1924; Phelps to Ryan, June 1924; Murray to Radford, January 1925.

105 Evelyn Phelps to Helen Ryan, Red Cross Files, January 2, 1924, RG 200, NAB II.

106 For works on black eugenic ideology and the ensuing interracial class tensions, see Dorr, *Segregation's Science*.

107 E. M. Murray to Pauline Radford, The American Red Cross, U.S. Veterans Hospital No. 91, Tuskegee, Alabama, December 3, 1924, RG 200, NAB II.

108 Ibid.

109 Linker, *War's Waste*, 9.

110 Natalia Molina, "Medicalizing the Mexican: Immigration, Race and Disability in the Early Twentieth Century U.S.," *Radical History Review* 94 (Winter 2006), 23, illustrates how cultural mores and legislative imperatives have coalesced around issues such as immigration—or in this case rehabilitation—"to write race on the body so indelibly that they are almost indistinguishable from biological inscription."

111 Stiker, *A History of Disability*, 123–125.

112 "War-Disabled Negroes in Training," 33.

113 Adas, *Dominance by Design*, 203.

114 Bender, "Perils of Degeneration," 7.

115 For analysis of race and racial hierarchies as "the changing same" in American political economy, see Sundiata Keita Cha-Jua, "The Changing Same: Black Racial Formation and Transformation as a Theory of the African American Experience," in Koditschek, Cha-Jua, and Neville, *Race Struggles*, 9–47.

CHAPTER 5. A NEW NEGRO TYPE

1 Ross, *The Origins of American Social Science*, 321.

2 Stoddard, *The Rising Tide of Color*; Bender, *American Abyss*, 236, 248–249.

3 Seitler, *Atavistic Tendencies*; Ross, *The Origins of American Social Science*, 322–323.

4 Committee on the American Negro Proposal, November 30, 1926, Davenport Papers, Box 12, Folder 2, APS.

5 Bender, *American Abyss*, 92–93.

6 Stoddard's notion of the "unfit" was not strictly confined to "the fecund colored races"—or the lower classes of whites, it also drew heavily on prewar notions of the white upper classes as "unfit" for abrogating their duty to propagate. Stoddard, *The Rising Tide of Color*, 220–221, 13–16; Baldwin, *Chicago's New Negroes*, 10–11.

7 Warren S. Thompson, "Race Suicide in the United States," *American Journal of Anthropology* 2, no. 1 (January–March 1919), 144–145; Gossett, *Race: The History of an Idea*, 340.

8 Ross, *The Origins of American Social Science*, 396–397.

9 Alain Locke, "The New Negro," in Holloway and Keppel, *Black Scholars on the Line*, 218.

10 Adas, *Dominance by Design*, 203.

11 Johnson, "Black Workers in the City"; A. E. Jenks to Hal Smith, Attn. Ford Motor Company, December 13, 1923, NRC Committee on Anthropology, Davenport Papers, Box 58, Folder 2, APS.

12 A. E. Jenks to Mr. Hal Smith, Attn. Ford Motor Company, December 13, 1923, NRC Committee on Anthropology, Davenport Papers, Box 58, Folder 2, APS.

13 Quoted in Lyman, *Militarism, Imperialism, and Racial Accommodation*, 16.

14 Adas, *Dominance by Design*, 281.

15 Johnson, "Black Workers and the City," 719.

16 John Richards "Some Experiences with Colored Soldiers," *Atlantic Monthly*, August 1919, 190. A common refrain among wartime evaluations of black soldiers was their supposed "childlike nature," and "affinity for the theatrics of drill and parade". Black soldiers were continually praised for their "imitative skill."

17 Johnson, "Black Workers and the City," 719.

18 Adas, *Dominance by Design*, 281; "The Determination of Racial Relationships by Means of Blood," *Journal of the American Medical Association* 73 (December 27, 1919), 1941–1942; DuBois, quoted in Lippman, "Negro Migration," *New Republic*, July 1, 1916.

19 Proposals for the Organization of Investigations of the American Negro, April 8, 1927, Davenport Papers, Series 1, Box 12, Folder 9, APS.

20 Boas to R. J. Terry, Committee on the American Negro, February 15, 1927, Boas Papers, APS; A. E. Jenks to Hal Smith, Attn. Ford Motor Company, December 13, 1923, NRC Committee on Anthropology, Davenport Papers, Series 1, Box 58, Folder 2, APS.

21 Baker, *Anthropology and the Racial Politics*, 5–7.

22 On anthropology as a function of imperial American health policies, see Warwick Anderson, "Pacific Crossings: Imperial Logics in U.S. Public Health Programs," in McCoy and Scarano, *Colonial Crucible*.

23 Boas to Terry, Committee on the American Negro, February 15, 1927, Boas Papers, Series 1, Box 71, Folder 6, APS.

24 Baker, *Anthropology and the Racial Politics*, 208–209; Lewis, *DuBois: Biography of a Race*, 351–353.

25 Baker, *Anthropology and the Racial Politics*, 25.

26 It was not until the 1930s and 40s that anthropology began to be used by activists, educators and lawyers to provide scientific proof of racial equality and better facilitate efforts towards desegregation. Baker, *From Savage to Negro*, 125–127.

27 These projects were united around a common concern regarding diversity's role as an agent of racial evolution (i.e., were "racial hybrids" a cause or effect of racial degeneration?). Report of the Committee on the Study of the American Negro, 1928, Davenport Papers, Series 1, Box 12, Folder 6, APS.

28 Ibid.

29 The crux of Grant's argument was the likeness between African Americans—seen as racially incompatible with whites—and immigrant groups tenuously seen as white. From a eugenic perspective, the "one drop rule" by which any degree of black ancestry made one a Negro, also applied to near-white races such as Jews. Slotkin, *Lost Battalions*, 455.

30 Ross, *The Origins of American Social Science*, 396–397.

31 Gershenhorn, *Melville Herskovits*, 124.

32 Wilson, *The Segregated Scholars*, 150–151; Ross, *The Origins of American Social Science*, 400.

33 Baker, *From Savage to Negro*, 93–94.

34 Stoddard, *Re-Forging America: The Story of Our Nationhood* (New York: Scribner, 1927), 256–257.

35 *New York Times*, October 27, 1921, 11.

36 Stoddard, *The Rising Tide of Color*, 220–222, 13–16; Bender, *American Abyss*, 248; Baldwin, *Chicago's New Negroes*, 10–11.

37 Guterl, *The Color of Race*, 216; Baldwin, *Chicago's New Negroes*, 10–11; Aptheker, *Pamphlets and Leaflets by W. E. B. DuBois*, 197.

38 On the riots as an agent of New Negro identity, see Barbara Foley, *Specters of 1919: Class and Nation in the Making of the New Negro* (Urbana: University of Illinois Press, 2008).

39 Albert Jenks, "University Training Course in Americanization or Applied Anthropology," *Scientific Monthly* (March 1921).

40 Charles Davenport, "Selecting Immigrants," (lecture, preliminary meeting on Human Migration, May 3, 1920), Davenport Papers, Series 1, Box 54, APS.

41 Suggestions for Committee on Scientific Problems of Human Migration (CSPHM), January 25, 1923; Report and Recommendations of the CSPHM, April 2, 1923; Proceedings of the Conference on Human Migration, November 18, 1922, Laura Spelman Rockefeller Memorial Foundation (LSRM) Collection, Box 21, Folder 5, Rockefeller Archive Center (RAC).

42 Bender, *American Abyss*, 40–41.

43 Frederick Hoffman, "Causes of Death in Primitive Races," Hoffman Papers, Box 29, Butler Library Rare Book and Manuscripts, Columbia University. See also Huntington, *Civilization and Climate*.

44 Davenport, *Lecture on Race Crossing* (n.d.), Davenport Papers, Series 1, Box 1, Folder 2, APS.

45 Francis A. Walker, "Restriction of Immigration," *Publications of the Immigration Restriction League* 33 (n.d.), 11.

46 Bender, *American Abyss*, 67.

47 Davenport, *Lecture on Race Crossing*.

48 Report and Recommendations of CSPHM, April 2, 1923; Report of CSPHM, July 1, 1923, LSRM Collection, Series 3, Box 58, Folder 629, RAC.

49 Ngai, *Impossible Subjects*, 23, 27.

50 Slotkin, *Lost Battalions*, 455; Stoddard, quoted in Bender, *American Abyss*, 244.

51 Matthew Jacobson notes that the "idea of a Caucasian race represents whiteness ratcheted up to a new epistemological realm of certainty." Jacobson, *Whiteness of a Different Color*, 94.

52 W. E. B. DuBois, "Brothers Come North," *The Crisis* 19 (January 1920), 105–106.

53 "Some Answers to Our Employment Problems," NUL Industrial Relations Department, Records of the NUL, Box 1, D7, LOC; "Will Negroes Stay in Industry?," *Survey* (December 14, 1918), 348–349.

54 Franklin, *From Slavery to Freedom*, 466–467; Slotkin, *Lost Battalions*; 405–407.

55 *Literary Digest*, June 14, 1919, 23.

56 James Weldon Johnson, "The Changing Status of Negro Labor," *National Conference on Social Welfare Proceedings* (1918), 383–384; *Eugenical News*, Fall 1925, 3.

57 Kelly Miller, "Negro Labor for the Steel Mills," *World's Work* (May–October 1923), 243.

58 "The American Negro as Soldier," *Literary Digest*, June 27, 1925, 14–15; "NUL: An Idea Made Practical," January 1920, General Education Board Collection, Series 1, Box 276, Folder 2876, 2–14, RAC.

59 At mid-century, 96% of black northerners and 90% of black westerners lived in urban areas. Berlin, *The Making of African Americans*, 154.

60 Johnson, "Black Workers and the City," 641.

61 Lyman, *Militarism, Imperialism, and Racial Accommodation*, 116–117.

62 Grant, *The Passing of the Great Race*, 209.

63 Thompson, "Race Suicide in the United States," 145.

64 Wilson, *The Segregated Scholars*, 23–25.

65 Trotter, *The Great Migration in Historical Perspective*, 5; "Helping Negroes Help Themselves" *Survey* 38, no. 12 (June 23, 1917), 278.

66 Gershenhorn, *Melville Herskovits*, 111. See Franz Boas, *The History of the American Race* (1912), *Report on an Anthropometric Investigation of the Population of the United States* (1922), *Race, Language, and Culture* (1940).

67 Wilson, *The Segregated Scholars*, 152–153.

68 Johnson, "Black Workers and the City"; Locke, "The New Negro," 218.

69 "Is Nature Solving the Negro Problem?," *World's Work* 47, (November–April 1923–1924), 131–132; William Pickens, "Migrating to Fuller Life," *Forum* 72 (November 1924), 593–607.

70 Mitchell, *Righteous Propagation*, 232–233.

71 Program of the *Eighth Annual Negro Health Week*, April 2–8, 1922, Records of the NUL, Part 1, Box 1, D5, LOC.

72 Toure Reed, "The Educational Alliance and the Urban League in New York: Ethnic Elites and the Politics of Americanization and Racial Uplift, 1903–1932," in Reed and Warren, *Reviewing Black Intellectual History*, 113–114.

73 Bender, *American Abyss*, 242–243; Mitchell, *Righteous Propagation*, 232–233; Carter Woodson, "The Exodus during the World War," in Holloway and Keppel, *Black Scholars on the Line*, 98.

74 Johnson, "Black Workers and the City," 643.

75 Mitchell, *Righteous Propagation*, 232–233.

76 Bay, *The White Image in the Black Mind*, 188–189.

77 Guterl, *The Color of Race*, 132–133.

78 A. E. Jenks to C. Davenport, Suggested Research Projects of the Committee on Race Characters (CRC), October 23, 1923, Davenport Papers, Series 1, Box 97, Folder 3, APS.

79 On the racial and gendered imperatives of imperial labor, see Bender and Lipman, *Making the Empire Work*.

80 Proposals for the Organization of Investigations of the American Negro, March 19, 1927, Davenport Papers, Series 1, Box 12, Folder 9, APS.

81 William McDougall, a British member of the NRC argued that America's "tardy" entry into the war—"if it was right in 1917 then it was equally right in 1914"—was due to the, "lack of a national mind; chiefly because it comprised many individuals and some large communities that were only very partially assimilated." William McDougall, "The Problems of Unassimilated Groups," *Proceedings of the Conference on Human Migration*, November 18, 1922, LSRM Collection, Box 24, Folder 4, RAC.

82 Committee on the American Negro, March 19, 1927, Davenport Papers, Series 1, Box 12, Folder 2, APS.

83 Ibid.

84 Garvey to Davis, October, 4, 1923, General Records of the Department of Labor, Files 1907–1942, RG 174, Box 19, Entry 1, NAB; Mitchell, *Righteous Propagation*, 230.

85 Carter, *The United States of United Races*, 124–125.

86 DuBois produced a striking visual rejoinder to these characterizations through a photographic exhibit of "respectable Negro types" at the 1900 Paris Exposition, later published as *Health and Physique of the American Negro* (1906). See Smith, *Photography on the Color Line*; Degler, *In Search of Human Nature*, 79–81.

87 Mitchell, *Righteous Propagation*, 203; Charles Johnson, "The Vanishing Mulatto," *Opportunity* (October 1925), 291.

88 Gershenhorn, *Melville Herskovits*, 40–45; Committee on the American Negro (Conference on Racial Differences, Washington, DC, February 1928), Davenport Papers, Series 1, Box 12, Folder 17, APS.

89 Davenport and Steggerda, *Race Crossing in Jamaica*, 125.

90 Park, quoted in Davenport and Steggerda, *Race Crossing in Jamaica*, 126.

91 McNeil, *Sex and Race in the Black Atlantic*, 3–4; Carter, *The United States of United Races*, 125.

92 Patterson to Yerkes, May 20, 1924, Davenport Papers, Series 1, Box 97, Folder 6, APS; Fish, "Human Engineering," 262.

93 Gould, *The Mismeasure of Man*, 228–229.

94 Committee on the American Negro (Conference on Racial Differences, Washington, DC, February 1928), Davenport Papers, Series 1, Box 12, Folder 17, APS.

95 Brantlinger, *Dark Vanishings*, 5–7; E. Fischer, *Die Rehobother Bastards und das Bastardierungsproblem beim Menschen* (*Rehobother Bastards and the Problem of Miscegenation Among Humans*) (1913), vii, 327; Louis R. Sullivan, "Anthropometry of the Siouan Tribes," *Anthropological Papers of the American Museum of Natural History* 23, no. 3 (1920), 81–174; Clark Wissler, "Distribution of Stature in the United States," *Scientific Monthly* 18, no. 2 (1924), 129–143; E. Rodenwaldt, *Die Mestizen auf Kisar* (*Mestizos of Kisar*) 2 vols. (Batavia, 1927); Leslie Dunn, *An Anthropometric Study of Hawaiians of Pure and Mixed Blood*, Vol. 11, Papers, Peabody Museum of American Archaeology and Ethnology (1928); Charles Davenport and Morris Steggerda, *Race Crossing in Jamaica* (Washington, DC: Carnegie Institute, 1929).

96 Davenport and Steggerda, *Race Crossing in Jamaica*, 51.

97 Davenport, "The Effects of Race Intermingling," *Proceedings of American Philosophical Society* 56, no. 4 (1922), 366. On racial admixture as a function and or impediment to the labor process, see Roediger and Esch, *The Production of Difference*, 108–121.

98 Davenport, "The Effects of Race Intermingling," 365.

99 Hoffman, "Race Pathology"; Davenport and Steggerda, *Race Crossing in Jamaica*, 3.

100 Years prior, in *Race Traits and Tendencies of the American Negro*, Hoffman had argued that "the difference in mortality of the two races is not so pronounced as between the white and colored populations of American cities, but is sufficiently

large to establish substantially the same *race tendency* to premature death among the colored population of the West Indies that we meet with among the colored population of this country." Hoffman, *Race Traits*, 69–71.

101 NRC Committee on Anthropology (Conference on Racial Differences, Washington, DC, Fall 1928), NRC Collection, APS; Davenport and Steggerda, *Race Crossing in Jamaica*, 289–291.

102 Davenport and Steggerda, *Race Crossing in Jamaica*, 289–291.

103 Ibid., 290–292.

104 Ibid.

105 Ibid.

106 Ibid., 464.

107 Emerson to Davenport, October 1927, Davenport Papers, Series 1, Box 36, Folder 22, APS.

108 Ibid.

109 Davenport and Steggerda, *Race Crossing in Jamaica (draft)*, November 8, 1928, Davenport Papers, Series 1, Box 27, Folder 2, APS.

110 Davenport, "Is the Crossing of Races Useful?," November 1929, Davenport Papers, Series 1, Box 26, Folder 8, APS.

111 Harris, "The Economic Foundations of American Race Division," in Holloway and Keppel, *Black Scholars on the Line*, 276–277.

112 Painter, *History of White People*, 274, 323.

113 Roediger and Esch, *The Production of Difference*, 186–193; Bender, *American Abyss*, 242–244.

114 Report and Recommendations of CSPHM, April 2, 1923; Report of CSPHM, July 1, 1923, LSRM Collection, Series 3, Box 58, Folder 629, RAC.

115 Jacobson, *Whiteness of a Different Color*, 94.

116 Gershenhorn, *Melville Herskovits*, 48.

117 U. G. Weatherly, "The West Indies as a Sociological Laboratory," *American Journal of Sociology* 29, no. 3 (November 1923), 290–304.

118 Gershenhorn, *Melville Herskovits*, 48.

119 Terry to Davenport, March 1929; Terry to Boas, March 1929, Preliminary Report of the Committee on the American Negro, 1926, Davenport Papers, Series 1, Box 18, Folder 17, APS.

120 Stoddard, *The Rising Tide of Color*, 120.

121 Otto Klienberg, *Negro Intelligence and Selective Migration* (New York: Columbia University Press, 1935), 60–62.

122 W. E. B. DuBois, "The African Roots of War," *Atlantic Monthly* 115, no. 5 (May 1915), 707–714.

123 Thompson, *The Making of the English Working Class*, 9.

124 Abram Harris, "The Economic Foundations of American Race Division," in *Black Scholars on the Line*, 288; Ross, *The Origins of American Social Science*, 472.

125 Locke, "The New Negro," 215.

126 Baker, *From Savage to Negro*, 127; Baker, *Anthropology and the Racial Politics*, 2–15.

EPILOGUE

1 Rabinbach, *The Human Motor*, 4–8; Roediger and Esch, *The Production of Difference*, 3–6.

2 Miles, *Capitalism and Unfree Labor*, 5–7; Paul Gilroy, "One Nation under a Groove," in Goldberg, *The Anatomy of Racism*, 265.

3 Lipsitz, *The Possessive Investment in Whiteness*, 177–179. Paul Gilroy charts the beginnings of modern rationality in the slave labor sugar plantations of the Caribbean. For Gilroy, the emergence of "race" in this early modern context is "an important reminder that making politics aesthetic was not a governmental strategy that originated in twentieth century fascism." Gilroy, *Black Atlantic*, 2, 47–49; Gilroy, *Against Race*, 56.

4 Wright, *White Man Listen!*, 72, 80.

5 Myrdal, quoted in Guterl, *The Color of Race*, 184. Myrdal observed that black social scientists—unlike their white counterparts—had long advocated social rather than biological models of race and racial difference and were therefore infinitely more modern than their white peers.

6 Cha-Jua, "The Changing Same," x, 38.

7 David Roediger, "White Without End? The Abolition of Whiteness; or the Rearticulation of Race," in Koditschek, Cha-Jua, and Neville, *Race Struggles*, 98; Cornel West and Jorge Klor de Alva, "On Black Brown Relations," in West, *The Cornel West Reader*, 499–514.

8 Cha-Jua, "The Changing Same," x, 38.

9 Singh, *Black Is a Country*.

10 Koditschek, Cha-Jua, and Neville, *Race Struggles*, x.

11 Michelle Alexander, *The New Jim Crow: Mass Incarceration in the Age of Color Blindness* (New York: New Press, 2012), 4–8. Alexander also notes that the United States imprisons a larger percentage of its black population than South Africa did at the height of apartheid.

12 Roediger, *How Race Survived U.S. History*, 210–211; Wayne, *Imagining Black America*, 45–47; Muhammad, *The Condemnation of Blackness*, 276–277. On the persistence of conceptions of black bodies as pathological or problematic in modern America, see Omi and Winant, *Racial Formation in the United States*; Bonilla-Silva, *Racism Without Racists*.

13 Miller, "Review of *Race Traits*" (emphasis mine).

14 On capitalism's classificatory impulse regarding the racial and or colonial subject, see Fanon, *The Wretched of the Earth*, 236–237.

15 Charles Blow, "Beyond 'Black Lives Matter,'" *New York Times*, February 9, 2015, http://nytimes.com/2015/02/09/opinion/charles-blow-beyond-black-lives-matter.html?.

16 Lipsitz, *The Possessive Investment in Whiteness*, 177; Bender, *American Abyss*, 247–256.

BIBLIOGRAPHY

MANUSCRIPT AND ARCHIVAL SOURCES

American Philosophical Society, Philadelphia, PA (APS)
National Research Council (NRC) Collection
Committee on Anthropology Collection
Charles B. Davenport Papers
Franz Boas Papers
Eugenics Records Office Records

University of Pennsylvania Archives, Philadelphia, PA
Eadweard Muybridge Papers

Rockefeller Archive Center, Sleepy Hollow, NY (RAC)
Laura Spelman Rockefeller Memorial (LSRM) Collection
General Education Board Collection
Social Science Research Council (SSRC) Collection

Schomburg Center for Research in Black Culture, Rare Book and Manuscripts,
New York Public Library (NYPL), New York, NY
Berry Family Letters

Butler Library Rare Book and Manuscripts, Columbia University, New York, NY
Frederick L. Hoffman Papers

National Archives Building, Washington, DC (NAB)
Record Group 15: Records of the Department of Veterans Affairs
Record Group 120: American Expeditionary Force

National Archives Building, College Park, MD (NAB II)
Record Group 112: Records of the Office of the Surgeon General (Army)
Record Group 163: Records of the Selective Service System
Record Group 165: Records of the War Department General and Special Staffs
Record Group 179: Records of the War Production Board
Record Group 200: Records of the Red Cross
Record Group 407: Records of the Adjutant General's Office

National Archives and Record Administration–Mid-Atlantic Region,
Philadelphia, PA
Record Group 32: Records of the U.S. Shipping Board

Library of Congress, Manuscript Division, Washington, DC (LOC)
Records of the National Association for the Advancement of Colored People (NAACP)
Records of the National Urban League (NUL)

GOVERNMENT DOCUMENTS (AUTHORED BY COMMITTEE)
Federal Board of Vocational Education, "What Every Disabled Soldier and Sailor
    Should Know," Washington, DC: Government Printing Office, November 1918.
Office of the Provost Marshal General, *Standards of Physical Examination Governing
    the Entrance to All Branches of the Armies of the United States*, Form 75, Washing-
    ton, DC: Government Printing Office, April 1918.
Office of the Provost Marshal General, *Manual of Instructions for Medical Advisory
    Boards*, Form 64, Washington, DC: Government Printing Office, 1918.
Office of the Provost Marshal General, *Manual of Instructions for Medical Advisory
    Boards*, Form 65, Washington, DC: Government Printing Office, 1918.
Office of the Surgeon General, *The Medical Department of the United States Army in
    the World War*, Vol. 13, Pt. 1, *Physical Reconstruction and Vocational Education*,
    Washington, DC: Government Printing Office, 1921.

PROCEEDINGS
Proceedings of The American Negro Academy
Proceedings of American Philosophical Society
Proceedings of the National Conference of Charities and Corrections
Proceedings of the National Conference on Social Welfare

UNPUBLISHED THESES
English, Daylanne Kathryn, "Eugenics, Modernism and the Harlem Renaissance,"
    Ph.D. diss., University of Virginia, 1996.
Perlman, Daniel, "Stirring the White Conscious: The Life of George E. Haynes," Ph.D.
    diss., New York University, 1972.

PUBLISHED PRIMARY AND SECONDARY SOURCES
Adas, Michael, *Dominance by Design: Technological Imperatives and America's Civiliz-
    ing Mission*, Cambridge, MA: Belknap Press of Harvard University Press, 2006.
Allen, Theodore, *The Invention of the White Race*, Vol. 1, *Racial Oppression and Social
    Control*, New York: Verso, 1994.
Anderson, Warwick, *Colonial Pathologies: American Tropical Medicine, Race, and
    Hygiene in the Philippines*, Durham, NC: Duke University Press, 2006.
Aptheker, Herbert, ed., *Pamphlets and Leaflets by W. E. B. DuBois*, New York: Kraus-
    Thomson, 1986.
Arneson, Eric, *The Black Worker: Race, Labor, and Civil Rights since Emancipation*,
    Urbana: University of Illinois Press, 2007.
Ayers, Edward, *The Promise of the New South: Life after Reconstruction*, New York:
    Oxford University Press, 1992.

Baker, Houston, *Turning South Again: Re-Thinking Modernism/Re-Reading Booker T. Washington*, Durham, NC: Duke University Press, 2001.

Baker, Lee, *Anthropology and the Racial Politics of Culture*, Durham, NC: Duke University Press, 2010.

——, *From Savage to Negro: Anthropology and the Construction of Race, 1896–1954*, Berkeley: University of California Press, 1998.

Baldwin, Davarian, *Chicago's New Negroes: Modernity, the Great Migration, and Black Urban Life*, Chapel Hill: University of North Carolina Press, 2007.

Baldwin, Davarian, and Minkah Makalani, eds., *Escape from New York: The New Negro Renaissance beyond Harlem*, Minneapolis: University of Minnesota Press, 2013.

Banta, Martha, *Taylored Lives: Narrative Productions in the Age of Taylor, Veblen, and Ford*, Chicago: Chicago University Press, 1993.

Barbeau, Arthur, and Florette Henri, *The Unknown Soldiers: Black American Troops in World War One*, Philadelphia: Temple University Press, 1974.

Baron, Ava, "Masculinity, the Embodied Male Worker, and the Historians Gaze," *International Labor and Working Class History* 69 (Spring 2006).

——, *Work Engendered: Toward a New History of American Labor*, Ithaca, NY: Cornell University Press, 1991.

Barringer, Paul, *The American Negro: His Past and Future*, Raleigh, 1900.

Bay, Mia, *The White Image in the Black Mind: African American Ideas about White People*, New York: Oxford University Press, 2000.

Bederman, Gail, *Manliness and Civilization: A Cultural History of Gender and Race in the United States, 1880–1917*, Chicago: University of Chicago Press, 1995.

Bender, Daniel, *American Abyss: Savagery and Civilization in the Age of Industry*, Ithaca, NY: Cornell University Press, 2009.

——, "Perils of Degeneration: Reform, The Savage Immigrant, and the Survival of the Unfit," *Journal of Social History* (Fall 2008).

——, *Sweated Work, Weak Bodies: Anti-Sweatshop Campaigns and Languages of Labor*, New Brunswick, NJ: Rutgers University Press, 2004.

Bender, Daniel, and Jana Lipman, eds., *Making the Empire Work: Labor and United States Imperialism*, New York: New York University Press, 2015.

Bendix, Reinhard, *Work and Authority in Industry: Ideologies of Management in the Course of Industrialization*, New York: Harper and Row, 1963.

Berlin, Ira, *The Making of African America: The Four Great Migrations*, New York: Viking, 2010.

——, *Many Thousands Gone: The First Two Centuries of Slavery in North America*, Cambridge, MA: Belknap Press of Harvard University Press, 1998.

Berman, Marshall, *All That Is Solid Melts into Air: The Experience of Modernity*, New York: Simon and Schuster, 1982.

Biddle, Tami Davis, "Military History, Democracy, and the Role of the Academy," *Journal of American History* 93 (March 2007).

Blackmon, Douglas, *Slavery by Another Name: The Re-Enslavement of Black Americans from the Civil War to World War II*, New York: Anchor Books, 2008.

Blake, Casey, *Beloved Community: The Cultural Criticism of Randolph Bourne, Van Wyck Brooks, Waldo Frank, and Lewis Mumford*, Chapel Hill: University of North Carolina Press, 1990.

Boas, Franz, *Anthropology and Modern Life*, New York: Norton, 1928.

———, *The Mind of Primitive Man*, New York: Macmillan, 1911.

Bonilla-Silva, Eduardo, *Racism without Racists: Color Blind Racism and the Persistence of Racial Inequality in America*, 3rd ed., New York: Rowman and Littlefield, 2009.

Boris, Eileen, "'You Wouldn't Want One of 'Em Dancing with Your Wife': Racialized Bodies on the Job in World War II," *American Quarterly* 50, no. 1 (1998).

Boris, Eileen, and Ava Baron, "The Body as a Useful Category for Working Class History," *Labor: Studies in Working Class History of the Americas* (Summer 2007).

Bourke, Joanna, *Dismembering the Male: Men's Bodies, Britain, and the Great War*, London: Reaktion, 1996.

Brantlinger, Patrick, *Dark Vanishings: Discourse on the Extinction of Primitive Races, 1800–1930*, Ithaca, NY: Cornell University Press, 2003.

Braun, Lundy, "Spirometry, Measurement, and Race in the Nineteenth Century," *Journal of the History of Medicine and Allied Sciences* 60, no. 2 (2005).

Braverman, Harry, *Labor and Monopoly Capital: The Degradation of Work in the Twentieth Century*, New York: Monthly Review Press, 1974.

Breisach, Ernst, *American Progressive History: An Experiment in Modernization*, Chicago: University of Chicago Press, 1993.

Bristow, Nancy, *Making Men Moral: Social Engineering during the Great War*, New York: New York University Press, 1996.

Brown, Elspeth, "Racialising the Virile Body: Eadweard Muybridge's Locomotion Studies, 1883–1887," *Gender and History* 17, no. 3 (November 2005): 627–656.

Bruinius, Harry, *Better for All the World: The Secret History of Forced Sterilization and America's Quest for Racial Purity*, New York: Knopf, 2006.

Buckley, Gail, *American Patriots: The Story of Blacks in the Military from the Revolution to Desert Storm*, New York: Random House, 2001.

Budreau, Lisa, *Bodies of War: World War One and the Politics of Commemoration in America, 1919–1933*, New York: New York University Press, 2010.

Burch, Susan, and Michael Rembis, eds., *Disability Histories*, Urbana: University of Illinois Press, 2014.

Campbell, James, ed., *Race, Nation, and Empire in American History*, Chapel Hill: University of North Carolina Press, 2007.

Carby, Hazel, *Race Men*, Cambridge, MA: Harvard University Press, 1998.

Carter, Greg, *The United States of the United Races: A Utopian History of Racial Mixing*, New York: New York University Press, 2013.

Cell, John, *The Highest Stage of White Supremacy: The Origins of Segregation in South Africa and the U.S. South*, Cambridge: Cambridge University Press, 1982.

Cobb, James, *The Most Southern Place on Earth*, New York: Oxford University Press, 1993.

Cohen, Deborah, *The War Come Home: Disabled Veterans in Britain and Germany, 1914–1939*, Berkeley: University of California Press, 2001.

Currell, Susan, and Christina Cogdell, eds., *Popular Eugenics: National Efficiency and Mass Culture in the 1930s*, Athens: Ohio University Press, 2006.

Davenport, Charles, and Albert Love, *Army Anthropology*, Washington, DC: Government Printing Office, 1921.

———, *Physical Examination of the First Million Draft Recruits*, Washington, DC: Government Printing Office, 1919.

Dawley, Alan, *Changing the World: American Progressives in War and Revolution*, Princeton, NJ: Princeton University Press, 2003.

Degler, Carl, *In Search of Human Nature: The Decline and Revival of Darwinism in American Social Thought*, New York: Oxford University Press, 1991.

Deloria, Phil, *Playing Indian*, New Haven, CT: Yale University Press, 1998.

Dinerstein, Joel, *Swinging the Machine: Modernity, Technology, and African American Culture between the World Wars*, Amherst: University of Massachusetts Press, 2003.

Dorr, Greg, *Segregation's Science: Eugenics and Society in Virginia*, Charlottesville: University of Virginia Press, 2008.

Downs, Jim, *Sick from Freedom: African American Illness and Suffering during the Civil War and Reconstruction*, Oxford: Oxford University Press, 2012.

DuBois, W. E. B., *Darkwater: Voices from within the Veil*, 1920. Reprint, New York: Washington Square Press, 2004.

———, *Health and Physique of the Negro American*, 1906.

———, *The Philadelphia Negro: A Social Study*, 1899. Reprint, Philadelphia: University of Pennsylvania Press, 1996.

———, *The Souls of Black Folk*, 1903. Reprinted in *Three Negro Classics*, New York: Avon, 1965.

Eggleston, Edward, *The Ultimate Solution to the American Negro Problem*, New York: AMS Press, 1913.

Ellison, Ralph, *The Collected Essays of Ralph Ellison*. Edited by John Callahan, New York: Modern Library, 2003.

Engels, Frederich, "The Condition of the Working Class in England, 1844." In *Marx-Engels Reader*, edited by Robert Tucker, New York: Norton, 1978.

Engs, Robert, *Educating the Disenfranchised and Disinherited: Samuel Chapman Armstrong and Hampton Institute, 1839–1893*, Knoxville: University of Tennessee Press, 1999.

Fabian, Ann, *The Skull Collectors: Race, Science, and America's Unburied Dead*, Chicago: University of Chicago Press, 2010.

Fanon, Frantz, *The Wretched of the Earth*, New York: Grove Press, 2004.

Farland, Maria, "W. E. B. DuBois, Anthropometric Science, and the Limits of Racial Uplift," *American Quarterly* 58, no. 4 (2006): 1017–1044.

Fields, Barbara, "Ideology and Race in American History." In *Region, Race, and Reconstruction: Essays in Honor of C. Vann Woodword*, edited by J. Morgan Kousser and James M. McPherson, New York: Oxford University Press, 1982.

Fioramonti, Lorenzo, *How Numbers Rule the World: The Use and Abuse of Statistics in Global Politics*, London: Zed Books, 2014.

Foucault, Michel, *The History of Sexuality*, Vol. 1, *An Introduction*. Translated by Robert Hurley, New York: Vintage, 1990.

Frader, Laura, "From Muscles to Nerves: Gender, 'Race,' and the Body at Work in France, 1919–1939," *International Review of Social History* 44 (1991).

Franklin, John Hope, *From Slavery to Freedom: A History of Negro Americans*, New York: Vintage, 1969.

Frederickson, George, *The Black Image in the White Mind: The Debate on Afro-American Character and Destiny*, Middletown, CT: Wesleyan University Press, 1987.

———, *White Supremacy: A Comparative Study in American and South African History*, Oxford: Oxford University Press, 1981.

Fussell, Paul, *The Great War and Modern Memory*, London: Oxford University Press, 1977.

Gains, Kevin, *Uplifting the Race: Black Leadership, Politics, and Culture in the Twentieth Century*, Chapel Hill: University of North Carolina Press, 1996.

Galishoff, Stuart, "Germs Know No Color Line: Black Health and Public Policy in Atlanta, 1900–1918," *Journal of the History of Medicine and Allied Sciences* 40 (1985).

Gamble, Vanessa, *Making a Place for Ourselves: The Black Hospital Movement, 1920–1945*, New York: Oxford University Press, 1995.

Gelber, Scott, "A 'Hard-Boiled Order': The Reeducation of Disabled WWI Veterans in New York City," *Journal of Social History* 39, no. 1 (2005).

Gerber, David, ed., *Disabled Veterans in History*, Ann Arbor: University of Michigan Press, 2000.

Gershenhorn, Jerry, *Melville Herskovits and the Racial Politics of Knowledge*, Lincoln: University of Nebraska Press, 2004.

Gerstle, Gary, *American Crucible: Race and Nation in the Twentieth Century*, Princeton, NJ: Princeton University Press, 2001.

Giddings, Franklin, *The Principles of Sociology*, New York: MacMillan, 1896.

Gilman, Sander, *Difference and Pathology: Stereotypes of Sexuality, Race, and Madness*, Ithaca, NY: Cornell University Press, 1985.

Gilroy, Paul, *Against Race: Imagining Political Culture beyond the Color Line*, Cambridge, MA: Belknap Press of Harvard University Press, 2000.

———, *Black Atlantic: Modernity and Double Consciousness*, Cambridge, MA: Harvard University Press, 1993.

Glenn, Brian J., "Postmodernism: The Basis of Insurance," *Risk Management and Insurance Review* (2003).

Goldberg, David Theo, ed., *The Anatomy of Racism*, Minneapolis: University of Minnesota Press, 1990.

Gordon, Sarah, "Prestige, Professionalism and the Paradox of Eadweard Muybridge's 'Animal Locomotion' Nudes," *Pennsylvania Magazine of History and Biography* 130, no. 1 (January 2006).

Gossett, Thomas, *Race: The History of an Idea in America*, New York: Schocken Books, 1969.

Gottlieb, Peter, *Making Their Own Way: Southern Blacks' Migration to Pittsburgh, 1916–1930*, Urbana: University of Illinois Press, 1987.

Gould, Benjamin, *Investigations in the Military and Anthropological Statistics of American Soldiers*, 1869. Reprint, New York: Arno Press, 1979.

Gould, Stephen J., *The Mismeasure of Man*, New York: Norton, 1996.

Gramsci, Antonio, *The Antonio Gramsci Reader: Selected Writings, 1916–1935*. Edited by David Forgacs, New York: New York University Press, 2000.

Grant, Madison, *The Passing of the Great Race*, New York: Scribner, 1916.

Grossman, James, *Land of Hope: Chicago, Black Southerners, and the Great Migration*, Chicago: University of Chicago Press, 1989.

Guterl, Matt, *The Color of Race in America, 1900–1940*, Cambridge, MA: Harvard University Press, 2001.

Guterl, Matt, and Christine Skwiot, "Atlantic and Pacific Crossings: Race, Empire, and the 'Labor Problem' in the Late Nineteenth Century," *Radical History Review* 91 (2005).

Guzda, Henry, "Social Experiment of the Labor Department: The Division of Negro Economics," *The Public Historian* 4, no. 4 (Fall 1982).

Hacking, Ian, *The Social Construction of What?*, Cambridge, MA: Harvard University Press, 1999.

Hale, Grace Elizabeth, *Making Whiteness: The Culture of Segregation in the South, 1890–1940*, New York: Vintage, 1999.

Haller, John, *Outcasts from Evolution: Scientific Attitudes toward Race in America, 1859–1900*, Urbana: University of Illinois Press, 1971.

Halpern, Rick, and Jonathan Morris, eds., *American Exceptionalism? U.S. Working Class Formation in an International Context*, New York: St. Martin's Press, 1997.

Hawley, Ellis, *The Great War and the Search for a New Order: A History of the American People and Their Institutions, 1917–1933*, New York: Waveland Press, 1997.

Haynes, George, *The Negro at Work in New York City*, New York: Arno Press, 1968.

———, *The Negro at Work during the World War and during Reconstruction*, Washington, DC: Government Printing Office, 1921.

Herrnstein, Richard, and Charles Murray, *The Bell Curve: Intelligence and Class Structure in American Life*, New York: Simon and Schuster, 1996.

Hoffman, Beatrix, *The Wages of Sickness: The Politics of Health Insurance in Progressive America*, Chapel Hill: University of North Carolina Press, 2001.

Hoffman, Frederick L., *History of the Prudential Life Insurance Company of America, 1875–1900*, Newark, NJ: Prudential Press, 1900.

———, *Race Traits and Tendencies of the American Negro*, New York: American Economic Association, 1896.

———, "Vital Statistics of the Negro," *Arena* (April 1892).

Hofstadter, Richard, *Social Darwinism in American Thought, 1860–1915*, Philadelphia: University of Pennsylvania Press, 1948.

Holloway, Jonathan Scott, and Ben Keppel, eds., *Black Scholars on the Line: Race, Social Science, and American Thought in the Twentieth Century*, Notre Dame, IN: University of Notre Dame Press, 2007.

Holt, Thomas, *Children of Fire: A History of African Americans*, New York: Hill and Wang, 2010.

————, "Explaining Race in American History." In *Imagined Histories: American Historians Interpret the Past*, edited by Anthony Molho and Gordan Wood, Princeton, NJ: Princeton University Press, 1998.

Holt, William S., *The Federal Board of Vocational Education: Its History, Activities and Organization*, 1922. Reprint, BiblioBazaar, 2000.

Horn, David, *The Criminal Body: Lombroso and the Anatomy of Deviance*, New York: Routledge, 2003.

Hrdlicka, Ales, *Anthropometry*, Philadelphia: Wistar Institute of Anatomy and Biology, 1920.

Humphreys, Margaret, *Intensely Human: The Health of the Black Soldier in the American Civil War*, Baltimore, MD: Johns Hopkins University Press, 2008.

Huntington, Ellsworth, *Civilization and Climate*, New Haven, CT: Yale University Press, 1924.

Hussie, W. M., "How Forestry and Tree Culture Concern the Disabled Soldier," *American Forestry* 24 (December 1918).

Ignatiev, Noel, *How the Irish Became White*, New York: Routledge, 1996.

Ireland, M. W., Charles Davenport, and Albert Love, "Part One: Army Anthropology." In *Medical Department of the U.S. Army in the World War*, Vol. 15, *Statistics*. Washington, DC: Government Printing Office, 1921.

Jackson, Giles, *The Industrial History of the Negro Race of the United States*, New York: Freeport, 1971.

Jacobson, Matthew Frye, *Barbarian Virtues: The United States Encounters Foreign Peoples at Home and Abroad, 1876–1917*, New York: Hill and Wang, 2000.

————, *Whiteness of a Different Color: European Immigrants and the Alchemy of Race*, Cambridge, MA: Harvard University Press, 1998.

James, C. L. R., *C. L. R. James on the Negro Question*. Edited by Scott McLemee, Jackson: University of Mississippi Press, 1996.

Jefferson, Robert, "Enabled Courage: Race, Disability, and Black World War II Veterans in Postwar America," *Historian* 65, no. 5 (September 2003).

Johnson, Walter, *Soul by Soul: Life Inside the Antebellum Slave Market*, Cambridge, MA: Harvard University Press, 1999.

Jones, Jacqueline, *American Work: Four Centuries of Black and White Labor*, New York: W.W. Norton, 1999.

Jones, William, *The Tribe of Black Ulysses: African American Lumber Workers in the Jim Crow South*, Urbana: University of Illinois Press, 2005.

Jordan, David Starr, *War and Waste: A Series of Discussions of War and War Accessories*, Garden City, NJ: Doubleday, 1914.

Kasson, John, *Houdini, Tarzan, and the Perfect Man: The White Male Body and the Challenge of Modernity in America*, New York: Hill and Wang, 2001.

Katz, Michael, and Thomas Sugrue, eds., *W. E. B. DuBois, Race, and the City: The Philadelphia Negro and Its Legacy*, Philadelphia: University of Pennsylvania Press, 1998.

Keane, Jennifer, *Doughboys, the Great War, and the Remaking of America*, Baltimore, MD: John Hopkins Press, 2001.

Kelley, Robin, *Freedom Dreams: The Black Radical Imagination*, Boston: Beacon Press, 2002.

——, *Race Rebels: Culture, Politics, and the Black Working Class*, New York: Free Press, 1996.

Kelves, Daniel, *In the Name of Eugenics: Genetics and the Uses of Human Heredity*, Cambridge, MA: Harvard University Press, 1998.

Kennedy, David, *Over Here: The First World War and American Society*, New York: Oxford University Press, 1980.

Kite, Gilbert, *Cotton Fields No More: Southern Agriculture, 1865–1980*, Lexington: University of Kentucky Press, 1984.

Koditschek, Theodore, Sundiata Kieta Cha-Jua, and Helen A. Neville, eds., *Race Struggles*, Urbana: University of Illinois Press, 2009.

Kudlick, Catherine, "Disability History: Why We Need Another 'Other,'" *American Historical Review* 108, no. 3 (June 2003).

Lansing, Michael, "Salvaging the Manpower: Conservation, Manhood, and Disabled Veterans during World War One," *Environmental History* 14, no. 1 (January 2009).

Larson, Edward, *Sex, Race, and Science: Eugenics in the Deep South*, Baltimore, MD: John Hopkins Press, 1995.

Lawrie, Paul R. D., "'Mortality as the Life Story of a People': Frederick L. Hoffman and Actuarial Narratives of African American Extinction, 1896–1915," *Canadian Review of American Studies* 43, no. 3 (2013).

Lears, Jackson, *Rebirth of a Nation: The Making of Modern America, 1877–1920*, New York: HarperCollins, 2009.

LeConte, Joseph, "The Race Problem in the South." In *Man and the State: Studies in Applied Sociology*, New York, 1892.

Lemire, Elise, *Miscegenation: Making Race in America*, Philadelphia: University of Pennsylvania Press, 2002.

Leonard, Thomas C., "More Merciful and Not Less Effective: Eugenics and American Economics in the Progressive Era," *History of Political Economy* 35 no. 4 (2003).

Lewis, David Levering, *W. E. B. DuBois, 1868–1919: Biography of a Race*, Vol. 1, New York: Henry Holt, 1993.

Liechtenstein, Alex, *Twice the Work of Free Labor: The Political Economy of Convict Labor in the New South*, New York: Verso, 1996.

Lipsitz, George, *The Possessive Investment in Whiteness: How White People Profit from Identity Politics*, Philadelphia: Temple University Press, 1998.

Litwack, Leon, *Trouble in Mind: Black Southerners in the Age of Jim Crow*, New York: Knopf, 1998.

Livingstone, James, *Pragmatism and the Political Economy of Cultural Revolution, 1850–1940*, Chapel Hill: University of North Carolina Press, 1997.

Logan, Rayford, *The Betrayal of the Negro, from Rutherford B. Hayes to Woodrow Wilson*, New York: Collier Books, 1965.

Longmore, Paul, *Why I Burned My Book, and Other Essays on Disability*, Philadelphia: Temple University Press, 2003.

Longmore, Paul, and Lauri Umansky, eds., *The New Disability History: American Perspectives*, New York: New York University Press, 2001.

Lott, Eric, *Love and Theft: Blackface Minstrelsy and the American Working Class*, Oxford: Oxford University Press, 1995.

Lyman, Stanford, *Militarism, Imperialism, and Racial Accommodation: An Analysis and Interpretation of the Early Writings of Robert E. Park*, Fayetteville: University of Arkansas Press, 1992.

Martin, James, *The Saga of Hog's Island, And Other Essays in Inconvenient History*, New York: R. Myles, 1977.

Marx, Karl, "The Economic and Philosophic Manuscripts 1841." In *The Portable Karl Marx*, edited by Eugene Kamenka, London: Penguin, 1983.

McCallum, Jack, *Leonard Wood: Rough Rider, Surgeon, Architect of American Imperialism*, New York: New York University Press, 2006.

McCartin, Joseph, *Labor's Great War: The Struggle for Industrial Democracy and the Origins of Modern American Labor Relations, 1912–1921*, Chapel Hill: University of North Carolina Press, 1997.

McCoy, Alfred, and Francisco Scarano, eds., *Colonial Crucible: Empire in the Making of the Modern American State*, Madison: University of Wisconsin Press, 2009.

McKee, James, *Sociology and the Race Problem: The Failure of a Perspective*, Urbana: University of Illinois Press, 1993.

McNeil, Daniel, *Sex and Race in the Black Atlantic: Mulatto Devils and Multiracial Messiahs*, New York: Routledge, 2010.

Menand, Louis, *The Metaphysical Club: A Story of Ideas in America*, New York: Farrar, Straus and Giroux, 2002.

Miles, Robert, *Capitalism and Unfree Labor: Anomaly or Necessity?*, London: Tavistock, 1987.

Mitchell, Michelle, "'The Black Man's Burden': African Americans, Imperialism, and Notions of Racial Manhood, 1890–1910," *International Review of Social History* 44 (1999).

——, *Righteous Propagation: African Americans and the Politics of Racial Destiny after Reconstruction*, Chapel Hill: University of North Carolina Press, 2004.

Montgomery, David, *Workers' Control in America: Studies in the History of Work, Technology and Labor Struggles*, Cambridge: Cambridge University Press, 1980.

Muhammad, Khalil G., *The Condemnation of Blackness: Race, Crime, and the Making of Modern Urban America*, Cambridge, MA: Harvard University Press, 2010.

Murphy, Gretchen, *Shadowing the White Man's Burden: U.S. Imperialism and the Problem of the Color Line*, New York: New York University Press, 2010.

Myrdal, Gunnar, *An American Dilemma: The Negro Problem and Modern Democracy*, New York: Pantheon, 1962.

Nelson, Scott, *Steel Drivin' Man: John Henry, the Untold Story of an American Legend*, Oxford: Oxford University Press, 2006.

Ngai, Mae, *Impossible Subjects: Illegal Aliens and the Making of Modern America*, Princeton, NJ: Princeton University Press, 2004.

Nordau, Max, *Degeneration*. Translated from the second edition of the German work, Lincoln: University of Nebraska Press, 1993.

Norrell, Robert, *Up From History: The Life of Booker T. Washington*, Cambridge, MA: Belknap Press of Harvard University Press, 2009.

Nye, Robert, *Crime, Madness, and Politics in Modern France*, Princeton, NJ: Princeton University Press, 1984.

———, "Degeneration, Neurasthenia and the Culture of Sport in Belle Époque France," *Journal of Contemporary History* 17, no. 1 (January 1982).

O'Connor, Alice, *Poverty Knowledge: Social Science, Social Policy, and the Poor in Twentieth-Century U.S. History*, Princeton, NJ: Princeton University Press, 2001.

Omi, Michael, and Howard Winant, *Racial Formation in the United States*, 4th ed., New York: Routledge, 2014.

Ott, Katherine, *Fevered Lives: Tuberculosis in American Culture since 1870*, Cambridge, MA: Harvard University Press, 1996.

Painter, Nell Irvin, *The History of White People*, New York: W.W. Norton, 2010.

———, *Standing at Armageddon: The United States, 1877–1919*, New York: W.W. Norton, 2008.

Parris, Guichard, and Lester Brooks, *Blacks in the City: A History of the National Urban League*, Boston: Little Brown, 1971.

Phillips, Kimberly, *Alabama North: African American Migrants, Community, and Working Class Activism in Cleveland, 1915–1945*, Urbana: University of Illinois Press, 1999.

Pick, Daniel, *Faces of Degeneration: A European Disorder, 1848–1918*, Cambridge: Cambridge University Press, 1989.

Pickens, Donald, *Eugenics and the Progressives*, Nashville, TN: Vanderbilt University Press, 1968.

Porter, Theodore, "Statistical Utopianism in an Age of Aristocratic Efficiency." In "Science and Civil Society," special issue, *Osiris* 17 (2002).

———, *Trust in Numbers: The Pursuit of Objectivity in Science and Public Life*, Princeton, NJ: Princeton University Press, 1995.

Price, Matthew, "Lives and Limbs: Rehabilitation of Wounded Soldiers in the Aftermath of the Great War." In "Cultural and Technological Incubations of Fascism," supplement, *Stanford Humanities Review* 5 (December 1996).

Rabinbach, Anson, *The Human Motor: Energy, Fatigue, and the Origins of Modernity*, Berkeley: University of California Press, 1990.

Reed, Adolph, Jr., *Class Notes: Posing as Politics, and Other Thoughts on the American Scene*, New York: New Press, 2000.

———, *W. E. B. DuBois and American Political Thought: Fabianism and the Color Line*, New York: Oxford University Press, 1997.

Reed, Adolph, Jr., and Kenneth Warren, eds., *Reviewing Black Intellectual History: The Ideological and Material Foundations of African American Thought*, Boulder: Paradigm Publishers, 2010.

Reed, Toure, *Not Alms but Opportunity: The Urban League and the Politics of Racial Uplift, 1910–1950*, Chapel Hill: University of North Carolina Press, 2008.

Roberts, Samuel, *Infectious Fear: Politics, Disease, and the Health Effects of Segregation*, Chapel Hill: University of North Carolina Press, 2009.

Rodgers, Daniel, *Atlantic Crossings: Social Politics in a Progressive Age*, Cambridge, MA: Belknap Press of Harvard University Press, 1998.

———, *The Work Ethic in Industrial America, 1850–1920*, Chicago: University of Chicago Press, 1978.

Roediger, David, *Colored White: Transcending the Racial Past*, Berkeley: University of California Press, 2003.

———, *How Race Survived U.S. History: From Settlement and Slavery to the Obama Phenomena*, London: Verso, 2008.

———, *The Wages of Whiteness: Race and the Making of the American Working Class*, London: Verso, 1991.

———, *Working Towards Whiteness: How America's Immigrants Became White*, New York: Basic Books, 2005.

Roediger, David, and Elizabeth Esch, *The Production of Difference: Race and the Management of Labor in U.S. History*, Oxford: Oxford University Press, 2012.

Rosenberg, Charles, and Janet Golden, eds., *Framing Disease: Studies in Cultural History*, New Brunswick, NJ: Rutgers University Press, 1987.

Ross, Dorothy, *The Origins of American Social Science*, Cambridge: Cambridge University Press, 1991.

Ross, Marlon, *Manning the Race: Reforming Black Men in the Jim Crow Era*, New York: New York University Press, 2004.

Runstedtler, Theresa, *Jack Johnson, Rebel Sojourner: Boxing in the Shadow of the Global Color Line*, Berkeley: University of California Press, 2012.

Rydell, Robert, *All the World's a Fair: Visions of Empire at American International Expositions*, Chicago: University of Chicago Press, 1984.

Said, Edward, *Orientalism*, New York: Vintage, 1994.

Sappol, Michael, *A Traffic of Dead Bodies: Anatomy and Embodied Social Identity in Nineteenth-Century America*, Princeton, NJ: Princeton University Press 2002.

Saxton, Alexander, *The Rise and Fall of the White Republic: Class Politics and Mass Culture in Nineteenth-Century America*, New York: Verso, 1990.

Schiber, Harry N., and Jane Schiber, "The Wilson Administration and the Wartime Mobilization of Black Americans," *Labor History* 10 (Summer 1969).

Schilling, Chris, *The Body and Social Theory*, London: Sage, 1993.

Schneider, Mark, *We Return Fighting: Civil Rights in the Jazz Age*, Boston: Northeastern University Press, 2002.

Schweik, Susan, *The Ugly Laws: Disability in Public*, New York: New York University Press, 2009.

Scott, Daryl Michael, *Contempt and Pity: Social Policy and the Image of the Damaged Black Psyche*, Chapel Hill: University of North Carolina Press, 1997.

Scott, Emmett J., *The American Negro in the World War*, 1919.

Scott, Joan, *Gender and the Politics of History*, New York: Columbia University Press, 1999.

Seitler, Dana, *Atavistic Tendencies: The Culture of Science in American Modernity*, Minneapolis: University of Minnesota Press, 2008.

Shenk, Gerald, *"Work or Fight!": Race, Gender, and the Draft in World War One*, New York: Palgrave-MacMillan, 2005.

Singh, Nikhil P., *Black Is a Country: Race and the Unfinished Struggle for Democracy*, Cambridge, MA: Harvard University Press, 2004.

Skocpol, Theda, *Bringing the State Back In*, Cambridge: Cambridge University Press, 1985.

Slotkin, Richard, *Lost Battalions: The Great War and the Crisis of American Nationality*, New York: Henry Holt, 2005.

Smethurst, James, *The African American Roots of Modernism: From Reconstruction to the Harlem Renaissance*, Chapel Hill: University of North Carolina Press, 2011.

Smith, John David, "W. E. B. DuBois, Felix Von Luschan, and Racial Reform at the *Fin de Siècle*," *Amerikastudien* 47, no. 1 (2002).

Smith, Mark, *How Race Is Made: Slavery, Segregation and the Senses*, Chapel Hill: University of North Carolina Press, 2006.

——, *Mastered by the Clock: Time, Slavery, and Freedom in the American South*, Chapel Hill: University of North Carolina Press, 1997.

Smith, Shawn Michelle, *Photography on the Color Line: W. E. B. DuBois, Race, and Visual Culture*, Durham, NC: Duke University Press, 2004.

Smith, William B., *The Color Line: A Brief in Behalf of the Unborn*, New York, 1905.

Smithers, Gregory, *Science, Sexuality, and Race in the United States and Australia, 1780s–1890s*, New York: Routledge, 2009.

Spiro, Jonathan, *Defending the Master Race: Conservation, Eugenics, and the Legacy of Madison Grant*, Burlington: University of Vermont Press, 2009.

Stanfield, John, "The Negro Problem Within and Beyond the Institutional Nexus of Pre-World War One Sociology," *Phylon* 43, no. 3 (1982).

Stanley, Amy Dru, *From Bondage to Contract: Wage Labor, Marriage, and the Market in the Age of Slave Emancipation*, Cambridge: Cambridge University Press, 1998.

Stepan, Nancy, *The Idea of Race in Science: Great Britain, 1800–1960*, London: MacMillan Press, 1982.

Stern, Alexandra Minner, *Eugenic Nation: Faults and Frontiers of Better Breeding in Modern America*, Berkeley: University of California Press, 2005.

Stiker, Henri-Jacques, *A History of Disability*, Ann Arbor: University of Michigan Press, 1999.

Stoddard, Lothrop, *The Rising Tide of Color Against White World-Supremacy*, New York: Scribner, 1920.

Stoneley, Peter, "Young Men and the Symmetrical Life," *New England Quarterly* 87, no. 2 (June 2014).

Summers, Martin, *Manliness and its Discontents: The Black Middle Class and the Transformation of Masculinity, 1900–1930*, Chapel Hill: University of North Carolina Press, 2004.

Sypher, Francis, *The Rediscovered Prophet: Frederick L. Hoffman (1865–1946)*, http://www.cosmos-club.org/web/journals/2000/sypher.html.

Taylor, Frederick W., *The Principles of Scientific Management*, New York: Dover Publications, 1998.

Thompson, E. P., *Customs in Common: Studies in Traditional Popular Culture*, New York: New Press, 1993.

———, *The Making of the English Working Class*, New York: Vintage, 1966.

———, *Witness Against the Beast: William Blake and the Moral Law*, New York: Vintage, 1993.

Thurber, Evangeline, *Preliminary Checklist of the General Administrative Files of the Rehabilitation Division Created under the Federal Board for Vocational Education (1918–1921) and the United States Veterans' Bureau (1921–1928)*, Washington, DC: National Archives, 1944, 6–7.

Trotter, Joe William, Jr., *Black Milwaukee: The Making of an Industrial Proletariat, 1910–1945*, 2nd ed., Urbana: University of Illinois Press, 2007.

———, ed., *The Great Migration in Historical Perspective: New Dimensions of Race, Class, and Gender*, Bloomington: Indiana University Press, 1991.

Visweswaran, Kamela, "Race and the Culture of Anthropology," *American Anthropologist* 100, no. 1 (1998).

Wailoo, Keith, *Dying in the City of the Blues: Sickle Cell Anemia and the Politics of Race and Health*, Chapel Hill: University of North Carolina Press, 2001.

Warren, Christian, "Northern Chills, Southern Fevers: Race-Specific Mortality in American cities, 1730–1900," *Journal of Southern History* 63 (February 1997).

Way, Peter, "Class and the Common Soldier in the Seven Years' War," *Labor History* 44, no. 4 (December 2003).

———, "Rebellion of the Regulars: Working Soldiers and the Mutiny of 1763–1764," *William and Mary Quarterly* 57, no. 4 (October 2000).

Wayne, Michael, *Imagining Black America*, New Haven, CT: Yale University Press, 2014.

Weare, Walter, *Black Business in the New South: A Social History of the North Carolina Mutual Life Insurance Company*, Durham, NC: Duke University Press, 1993.

Weiss, Nancy, *The National Urban League, 1910–1940*, New York: Oxford University Press, 1972.

Wells, Ida B., *The Reason Why the Colored American Is Not in the World's Columbian Exposition*, Urbana: University of Illinois Press, 2000.

West, Cornel, *The American Evasion of Philosophy: A Genealogy of Pragmatism*, Madison: University of Wisconsin Press, 1989.

———, *The Cornel West Reader*. Edited by Cornel West, New York: Basic Books, 1999.

Wheen, Francis, *Marx's 'Das Kapital': A Biography*, Vancouver: Douglas and McIntyre, 2007.

Wilkerson, Isabel, *The Warmth of Other Suns: The Epic Story of America's Great Migration*, New York: Vintage, 2010.

Williams, Chad, *Torchbearers of Democracy: African America Soldiers in the World War One Era*, Chapel Hill: University of North Carolina Press, 2010.

Williams, Kidada E., *They Left Great Marks on Me: African American Testimonies of Racial Violence from Emancipation to World War One*. New York: New York University Press, 2012.

Williams, Raymond, *Marxism and Literature*, New York: Oxford University Press, 1978.

Williams, Vernon, *From a Caste to a Minority: Changing Attitudes of American Sociologists Toward Afro-Americans, 1896–1945*, New York: Greenwood Press, 1989.

Williamson, Joel, *New People: Miscegenation and Mulattoes in the United States*, New York: New York University Press, 1984.

Wilson, Francille, *The Segregated Scholars: Black Social Scientists and the Creation of Black Labor Studies*, Charlottesville: University of Virginia Press, 2006.

Winston, James, *Holding Aloft the Banner of Ethiopia: Caribbean Radicalism in Early Twentieth-Century America*, New York: Verso, 1999.

Witkowski, Jan, and John Inglis, eds., *Davenport's Dream: 21st Century Reflections on Heredity and Eugenics*, Cold Spring Harbor, NY: Cold Spring Harbor Laboratory Press, 2008.

Wolff, Megan, "The Myth of the Actuary: Life Insurance and Frederick Hoffman's Race Traits and Tendencies of the American Negro," *Public Health Reports* 121 (January–February 2006).

Wright, Richard, *White Man Listen!*, New York: Anchor Books, 1964.

Yerkes, Robert, *Psychological Examining in the U.S. Army*, Washington, DC: Government Printing Office, 1921.

Zelizer, Viviana, *Morals and Markets: The Development of Life Insurance in the United States*, New York: Columbia University Press, 1979.

Zimmerman, Andrew, *Alabama in Africa: Booker T. Washington, the German Empire, and the Globalization of the New South*, Princeton, NJ: Princeton University Press, 2010.

Zinn, Howard, and Anthony Arnove, *Voices of a People's History of the United States*, New York: Seven Stories Press, 2004.

INDEX

Actuarial science, 7–8, 12, 19, 36, 94; black criminality, 22; narratives of race suicide, 31, 33, 42; racial labor division, 14–16, 18, 30, 37–38, 41. *See also* Race suicide

African Americans, 10, 13, 43, 54, 55, 66, 122, 125, 128, 133, 157, 165, 169–170; actuarial definitions of (*see also* Great Migration), 18, 36; incarceration rates, 23, 26, 171–172; insurance companies, 33–34; "in the way peoples," (Baker), 42; participation in World War I, 82–93; postwar critiques of military service, 148–150; proletarianization of, 5–6, 14, 37, 41, 67, 70, 90, 105, 171; sexuality, 22, 62; support for and against war effort, 83; transition from bondage to contract, 14; wealth, 174n18

African American women: "daughters and mothers of race," 57; and sexuality, 25–26; wartime employment, 58–59

American Dilemma, An (Myrdal), 32, 170

American Expeditionary Force (AEF), 85, 147

American Federation of Labor (AFL), 60, 101

American Indians, 10, 42, 139–141; race suicide of, 7, 14, 19, 37

American Journal of Physical Anthropology, 96, 143

American Journal of Sociology, 42–43

American Nervousness (Beard), 106

American Red Cross, 123, 126, 128–131, 153

Anthropology: anthropometry and, 94; *Army Anthropology*, 95, 104–105; Committee on Anthropology (COA), 6, 9, 73, 94–95, 145; race and physical anthropology, 2, 73–74, 81, 88, 90, 92–93, 96, 101, 138, 140–141, 143, 158, 170

Anthropometry: in Civil War, 85, 93; postwar uses, 104–105; racial dynamics of, 9, 96, 108, 170, 177n47; sartorial aspects, 102–103. *See also* Anthropology; World War I

Armstrong, Samuel, 14–15, 178n4

Army Sanitary Corps, 92, 113

Atavism, 98, 132; racial dynamics of, 3, 13, 23, 35, 85, 135

Bailey, Ben, 1–3, 5; boxing career, 7, 180n34

Biology, 5, 37, 143, 145, 147; and crime, 151; racial dynamics of, 4, 10–11, 16, 20–22, 30, 42, 44, 133, 135, 140, 152, 165, 167, 174n17

Biometrics, 75

Black Lives Matter, 172

Black Worker in the City, The (Johnson), 152

Boas, Franz, 77, 140, 190n28

Boer War, 80

Bolshevism, 62, 144, 155

Bourne, Randolph, 71–72

Boxing, 1

Bureau of War Risk Insurance (BWRI) 67, 115, 118, 125

Canada: blacks in, 146; vocational rehabilitation, 114

ABOUT THE AUTHOR

Paul R. D. Lawrie is Assistant Professor of History at the University of Winnipeg, Canada.